LIBRARY
VT TECHNICAL COLLEGE
RANDOLPH CTR VT 05061

We are now in the 'Age of Environmental Enlightenment.' Global warming is seen as a major threat to the well-being of the world's communities. Fear abounds, but what does it all mean? Do the scientists know what is going on? If so what can be done?

Michael Glantz draws attention to the relationship between society and climate change. He examines the notion that 'drought follows the plow.' The latest speculation is that dry areas will get drier and wet regions wetter, so why are many communities being moved to these dry regions? With contributions from colleagues in the worst-hit areas of the world, Michael Glantz challenges decisionmakers to realize that human activity can bring about an increase in drought-related food production problems. This is a book for all those who want to know more about the interactions between climate and society. Policy-makers, scientists, and the general public worldwide need to read this book.

Drought follows the plow

Drought follows the plow:
cultivating marginal areas

Edited by
MICHAEL H. GLANTZ

Published by the Press Syndicate of the University of Cambridge
The Pitt Building, Trumpington Street, Cambridge CB2 1RP
40 West 20th Street, New York, NY 10011-4211, USA
10 Stamford Road, Oakleigh, Melbourne 3166, Australia

© Cambridge University Press 1994

First published 1994

Printed in Great Britain at the University Press, Cambridge

A catalogue record for this book is available from the British Library

Library of Congress cataloguing in publication data available

ISBN 0 521 44252 4 hardback

ISBN 0 521 47721 2 paperback

Dedication

This book is dedicated to Dirceu Murilo Pessoa, Secretary General of the Brazilian Ministry of Agriculture and Agrarian Reform, whose untimely death in June 1987 deprived us of a dedicated scholar and policymaker. During the nearly three decades of his professional life he focused his energies on addressing and solving problems of land equity, social justice, and development. Trained as an economist both in Brazil and France, he always sought to bring international attention to the plight of his native, drought-stricken Northeastern region of Brazil. Dirceu profoundly understood what it took to be a scholar with a role to play in solving his nation's problems.

Contents

Preface	xi
List of contributors	xvii
Introduction	
MICHAEL H. GLANTZ	1
PART I. Drought, desertification and food production	
MICHAEL H. GLANTZ	7
Drought	9
Desertification	12
Food production: the role of marginal lands	15
Cultivating the dry margins	21
PART II. Case studies and conclusions	
The West African Sahel	
MICHAEL H. GLANTZ	33
Somalia	
JÖRG JANZEN	45
The Brazilian Nordeste (Northeast)	
ANTONIO MAGALHÃES AND PENNIE MAGEE	59
The dry regions of Kenya	
DAVID CAMPBELL	77

Australia
    R. LES HEATHCOTE   91

Ethiopia
    JAMES McCANN   103

Northwest Africa
    WILL SWEARINGEN   117

The Virgin Lands Scheme in the former Soviet Union
    IGOR ZONN, MICHAEL H. GLANTZ AND ALVIN RUBINSTEIN   135

South Africa
    COLEEN VOGEL   151

Is the stork outrunning the plow?
    MICHAEL H. GLANTZ   171

Notes   177

Index   195

Preface

There is a good chance that future generations will look back at the 1980s as the beginning of the 'Age of Environmental Enlightenment.' In that decade worldwide interest in the environment greatly increased, an interest based on a large number of environmental changes increasingly perceived by the public as global threats.

During the 1970s, desertification, water resources, technology transfer, energy needs, food production, population increases, human settlements, and the human environment were topics of concern to the United Nations General Assembly. In the 1980s the focus of concern shifted to such global issues as ozone depletion, climate change, tropical deforestation, and biodiversity. These issues have since become the centerpieces of a wide-ranging global change research agenda. In 1987, the ozone depletion issue was seriously addressed by political leaders around the world with the signing of the Montreal Protocol.[1] Also in that year the World Commission on Environment and Development issued its report on sustainable development, entitled *Our Common Future*.[2]

A few years later, rich and poor nations around the globe had to deal with each other on environment and development issues at the United Nations Conference on Environment and Development (UNCED) held in Rio de Janeiro in June 1992. At this Earth Summit a climate framework convention was signed, as were conventions on biodiversity and on forest protection.

It seems that governments are beginning to listen to their citizens' pleas to protect the Earth. Nations with different and often conflicting interests have been prompted into cooperative action on regional and global scales in order to address, if not resolve, some of those problems perceived most important to them.

The collapse of communism added to pressures on governments everywhere by allowing the world to see how poorly the environment had been treated throughout the former Soviet Union and eastern Europe. Shocking photographs of children going to school in gas masks to avoid noxious fumes from polluting industrial plants were matched by horror stories about the aftermath of the Chernobyl nuclear disaster or about the adverse health and ecological effects of decades of cotton production in the Aral Basin in Central Asia.

Opinion surveys involving people in the street[3] as well as policymakers and scientists[4] in their offices support the contention that environmental degradation is on the minds of many. The media, too, have become increasingly aggressive in their coverage of major global as well as local environmental issues.

It seems as though the interest, if not the will, is there to do something about improving the quality of life by improving the quality of the environment. Some problems considered to be globally threatening are being addressed front and center, such as ozone depletion. Following the belated 'discovery' of the Antarctic ozone hole, the possibility of stratospheric ozone depletion over the Arctic region, and the linkage of ozone layer depletion with a heightened risk of skin cancer in the midlatitudes of the Northern Hemisphere, governments responded with an international protocol to end the use of chlorofluorocarbons (CFCs) in the next several years. Societies have also come to realize that many of today's environmental problems will probably be with us throughout the next century. For example, ozone-eating CFCs have a long residence time in the stratosphere (60 to 100 years), even if their usage were ended today.

The 1980s also witnessed a sharp increase in national and international political concern about the possible implications of a global warming of the atmosphere. Many scientific reports, journal papers, and popular articles on all sides of the issue have appeared on the scene, highlighting global warming as a potential threat to the well-being of the global community. Scenarios generated by general circulation models of the atmosphere suggest that such a global climate change could bring about serious but as yet unidentified shifts in climate regimes at the regional level. However, a considerable number of uncertainties continue to cloud the global warming issue: the rate of temperature change; the ability of human societies and ecosystems to adjust to those rates of change; cloud feedback mechanisms that could reverse the warming trend; the possibility that the anthropogenic enhancement of global warming is countering a natural movement toward an ice age; the level of reliability of scenarios generated by general circulation models of the atmosphere for policy making; and global warming implications for changes in the frequency, intensity, and duration of extreme meteorological events such as droughts, floods, and frosts.

The climate change issue has also drawn attention to the obvious as well as the subtle interactions between poverty of the environment and the environment of poverty. Calls to protect the quality of the Earth's atmosphere and, more generally, the Earth's environment for future generations to enjoy has raised ethical questions about what we are doing in this regard for those few generations of people alive today. In other words, concern about inter-generational equity issues has led to renewed calls for intra-generational equity. How to break this vicious downward spiral of the quality of life and environment at the regional and local level is a key concern.

Nevertheless, while there may be no overwhelming consensus as of today[5] on various aspects of the scientific theories concerning the global warming issue (other than agreement that human activities are adding increasing amounts of greenhouse gases to the atmosphere and that such gases can in theory generate a greenhouse effect), there is certainly enough evidence to alarm the global community into realizing that precautionary action is warranted. Such action can take many forms. One popular political action is to call for more research to reduce scientific uncertainties (for example, will the warming increase evaporation, which will increase the amount of cloud cover, which will lead in turn to a cooling of the atmosphere?). Another approach is to make decisions to prevent or at least slow down increases in the rate of emissions of greenhouse gases. In fact, there have been numerous suggestions about how to cope with the issue of a greenhouse-gases-induced global warming.[6]

With regard to the call for more research, the scientific community has for some years focused on early detection of global warming. Climate varies seasonally, interannually, and on decadal and longer time scales. Depending on the particular region of concern on the Earth's surface, climate variability can range from small to large. Precipitation in arid areas, for example, is skewed to dryness, with a few rainfall episodes that are far above average being balanced out by a larger number of below-average rainfall events; conditions that are statistically average seldom occur. In such regions rainfall variability is high, making it very difficult to identify a climate change signal in the midst of climate variability noise.

One attempt among several at early detection of climate change has been to look at a change in the frequency or intensity of extreme meteorological events. Droughts and floods, as major extreme natural hazards, have been singled out as possible indicators of changing climate. In the mid-1970s, increases in the number of meteorological stations in India reporting drought were suggested as an early indicator of the regional impacts of a global climate change (at that time scientists focused on the possibility of a global cooling).[7]

Today many researchers speculate that with global warming of the atmosphere there will probably be an increase in drought-related crop failures in areas already subject to drought. They contend that dry areas will get drier and wet areas wetter. Yet, observations show that people are moving into regions that are relatively less fruitful with respect to agricultural productivity than the lands from which they emigrated. The assumption here is that much of the best rain-fed agricultural land (where agriculture is directly dependent on rainfall) has already been put into production, and that by the movement of farmers into relatively less productive regions, the risk of crop failure would probably increase. This would especially be the case if the same crops that these migrants had grown in the more productive agricultural areas continue to be grown in the marginal areas. In addition, major alterations in the vegetative cover in marginal areas could change regional rainfall patterns as a result, for example, of changes in land surface properties such as albedo (land surface reflectivity). Thus, we are likely to hear more about crop failures in the future because of human impacts on these fragile environments, whether or not the global climate changes.

The notion that 'drought follows the plow' draws attention to the land-use changes that are taking place – for the worse. Although this book focuses on the drier regions of the globe, similar processes can occur in the high-rainfall areas as well. Inappropriate land use can reduce agricultural productivity for a variety of reasons and can increase the risk of agricultural and hydrological droughts and associated food shortages. Human activities in different societies around the globe may possibly confound our ability to detect correctly the onset of the projected global warming and its earliest societal and ecological impacts.

The notion that drought might follow the plow also underscores a problem with regard to the early detection of or perceptions about global warming. If the notion proves to be a valid one, it will no longer be a matter, for example, of counting the number of weather stations reporting meteorological droughts or crop monitoring stations reporting agricultural droughts in a given area to determine whether the frequency or intensity of extreme events has increased (a proposed sign of climate change). Droughts measured in terms of a decline in crop production or an outright crop failure will have to be carefully scrutinized so as to identify the cause correctly. Such scientific scrutiny will enable researchers and policymakers to determine which drought impacts might be blamed on physical processes and which might be blamed on socioeconomic factors.[8]

I first presented the notion that 'drought follows the plow' some years ago at an international conference on climate forecasting held in Fortaleza, the capital of the northeastern Brazilian state of Ceará. Following that presentation, a Brazilian economist commented that a similar process of land use was already under way in his country. I suggested that he prepare a section describing the process in his country for inclusion in the original paper. In time, researchers from other disciplines and other countries responded similarly to the notion and were also asked to prepare similar sections. This much-expanded publication is the result of that process.

As the initial author and, later, shepherd and editor of this manuscript, it was difficult to determine how many case studies from around the world would suffice. The cases chosen were selected to present an effective, reliable, and credible argument about the likelihood that a process of 'drought following the plow' could become a more frequent and widespread occurrence in future decades. This would occur because of a variety of irrefutable, measurable, continuing demographic changes, such as high population growth rates in developing areas (within industrialized as well as industrializing countries) and increases in the demand for land to produce basic food stuffs as well as intensified cash crop production.

The contibutors to this book hope that researchers familiar with other regional and local examples of the notion that drought follows the plow will undertake similar assessments. Such regional and local assessments will enable those researchers searching for regional signals of a global climate change to attribute appropriately to human factors or to climate change any unusual changes in the frequency, distribution, magnitude, and timing of droughts. A more realistic attribution would allow policymakers to better match drought problems and their causes with appropriate solutions to prevent, mitigate, or adapt to their consequences.

Boulder, Colorado
October 1993

Michael H. Glantz

Contributors

MICHAEL H. GLANTZ
Senior Scientist and Director of the Environmental and Societal Impacts Group, National Center for Atmospheric Research, PO Box 3000, Boulder, CO 80307, USA*

DAVID CAMPBELL
Department of Geography, Michigan State University, East Lansing, MI 48823, USA

R. LES HEATHCOTE
Flinders University of South Australia, GPO Box 2100, Adelaide 5001, Australia

JÖRG JANZEN
Freie Universität Berlin, FB 24/WE5, Grunewaldstrasse 5, D-1000 Berlin 41, Germany

ANTONIO MAGALHÃES
Esquel Brasil Foundation, Brasilia DF, Brazil

PENNIE MAGEE
Visiting Scientist, Environmental and Societal Impacts Group, National Center for Atmospheric Research, PO Box 3000, Boulder, CO 80307, USA*

*The National Center for Atmospheric Research is sponsored by the National Science Foundation

JAMES McCANN
African Studies Center, Boston University, 270 Bay State Road, Boston, MA 02215, USA

ALVIN Z. RUBINSTEIN
Department of Political Science, University of Pennsylvania, 217 Stiteler Hall, Philadelphia, PA 19104-6215, USA

WILL SWEARINGEN
Earth Sciences Department, Montana State University, Bozeman, MT 59717, USA

COLEEN VOGEL
Climatology Research Group, University of the Witwatersrand, Johannesburg, South Africa

IGOR ZONN
Russian Engineering Academy, Foreign Relations Office, Soyuzvodproject, 17005 Moscow, Russia

The photographs in each chapter were supplied by the contributor, unless otherwise noted.

Introduction

MICHAEL H. GLANTZ

To those who have studied the geography or the history of the US Great Plains, 'drought follows the plow' might appear to be a simple play on words. This notion turns around a nineteenth-century belief that human activities (primarily plowing the ground) could ultimately bring about an increase in rainfall in a given area. That belief became popularized as 'rain follows the plow.' In fact, it *is* a play on words, but it does accurately describe what we believe to be an important aspect of the interactions between human activities, land surfaces, and regional atmospheric processes occurring today.

The original notion that rain follows the plow was explicitly applied to the US Great Plains during the 1870s and 1880s in response to increases in regional rainfall that were perceived to have been human-induced because they accompanied the development of new human settlements. Those perceptions were based on the belief that human activities such as tree planting, land clearing, cultivation, and fires could alter atmospheric processes so as to increase local and regional rainfall. At the time, hypotheses in support of such a contention were plentiful in the scientific literature, as well as in the popular press.

Although the notion, and the scientific hypotheses derived to support it, faced challenges from the outset, the belief managed to persist. However, the credibility and reliability of the notion was soon challenged by Mother Nature, with the occurrence of severe droughts in the US Great Plains from the late 1880s to the 1890s. The return of good rains that followed this extreme drought period renewed the belief that rain follows the plow. That the belief still persists in some circles today is evidenced by the continual stream of scientific hypotheses in the form of research and popular articles and in actual engineering proposals about land-use changes made for the primary purpose of modifying atmospheric processes in order to enhance regional rainfall.[1]

As we approach the end of the twentieth century, there is enough evidence to suggest that the opposite process is now occurring around the globe. Drought appears to be following the plow in marginal land areas worldwide, as a direct result of the complex interactions between environmental and societal factors. This realization is based on the assumption that, generally speaking, most of the best rain-fed agricultural land is either already in production or has been devoted to other human activities. The problem stems from the diminishing availability of the resources that will be required to sustain increasing population numbers in marginal areas, e.g., land at the climatic margin (low, unreliable rainfall or short growing seasons), at the productive soil margin (poor, unstable soils), or at the topographic margin (unsuitable terrain). The use of the term 'marginal' is meant to convey the characteristic of being located at the fringe. Marginal in an agricultural sense refers to the notion that the risks associated with cultivating a given piece of land successfully and sustainably are quite high and that the chance of crop failure in any given year is also relatively high. An economist might view marginal lands as those where the prices received for the marketed goods barely cover the costs of production.

One might effectively argue that wherever people tend to move in future decades to open up new farmland will most likely be less productive in the long run under rain-fed agricultural conditions than the locale from which they emigrate. This argument is based on the premise that, if there were better land available for rain-fed agricultural production, that land would already have been put into cultivation, barring the existence of some geophysical, economic, political, or social constraint. Thus, as agriculturally marginal lands such as those in arid and semiarid areas are increasingly put under the plow, the probability that these new agricultural activities will be adversely affected by agricultural drought will be likely to increase, even under meteorological conditions considered to be normal for the area. Under conditions generated by a mismatch of social and environmental conditions, land degradation, even desertification, often occurs. Furthermore, the encroachment of agricultural activities into marginal areas is a process that may now be accelerating as a result of sharply increasing population pressures, the increases in the demand and need for nutrition, and the dwindling availability to farmers of regional and local natural resources.

The increase in the frequency or intensity of agricultural droughts in these marginal areas will not necessarily have been the result of changing precipitation patterns in the region but will have resulted from these inappropriate (with respect to the regional environmental conditions) agricultural practices. Thus, newly introduced agricultural production activities would be likely to be adversely affected by normal meteorological conditions. More often than not, technologies and crops will have been imported from relatively more productive neighboring areas and would require a greater water supply or higher soil fertility than such marginal areas can provide over the long term. For these reasons we challenge anew the lingering belief that 'rain follows the plow.' More importantly, we address the more ominous implications to the environment of the movement of populations into areas known to be marginal for sustainable agricultural production as regards climate, topography, or soil.

The distinction between human-induced droughts and those caused by natural climatic conditions becomes all the more important, given the widespread concern about the likelihood of human-induced global climate change. A key point here is that, because of social processes encompassed by the 'drought follows the plow' notion, we are likely to hear more about droughts in the future as populations increase, as social groups become marginalized (in a socioeconomic sense), and as individuals and governments search for new lands to cultivate.

There are constant suggestions that a global warming of the atmosphere could lead to an increase in the frequency of meteorological droughts where they presently occur and that they could be more intense, of longer duration and, by implication, more devastating to human activities. Such a scenario suggests that we may also be hearing more about droughts in the future because of the regional and local effects of a global climate change, if the present-day atmospheric fluctuations prove to have been part of a permanent global warming trend.

Thus, the impacts of increasing population pressures on marginal areas could easily confound our potential ability to detect a climate change in its early stages through the monitoring of the number of drought episodes over time in a given area. Population pressures, government bias toward cash crop as opposed to food crop production, and the dividing of land into smaller and smaller agricultural plots have each in its own way brought about changes in the types and locations of land-use activities. Such factors greatly affect land-surface processes, even in the absence of a climate change of a few degrees Celsius by the middle of the next century.

Each of these two processes – one social and the other physical – increases the likelihood of crop failures in marginal areas. Each needs to be explicitly identified and its contributions to regional crop failures and food shortages determined and acknowledged, if appropriate response strategies are to be devised.

Drawing on specific examples provided by researchers from around the globe, the 'drought follows the plow' notion is discussed in more detail in the following sections. Case studies include the West African Sahel, Somalia, Northeast Brazil, Kenya, Australia, Ethiopia, Northwest Africa, South Africa, and western Siberia and northern Kazakhstan (Figure 1). A brief discussion of the US Great Plains in the context of the original 'rain follows the plow' concept is also presented. Each case study raises questions about the adequacy of the way we look at, prepare for, and respond to contemporary droughts. Each case also raises concern about how we might respond to droughts in the future, regardless of whether they prove to have been human-induced or naturally occurring phenomena.

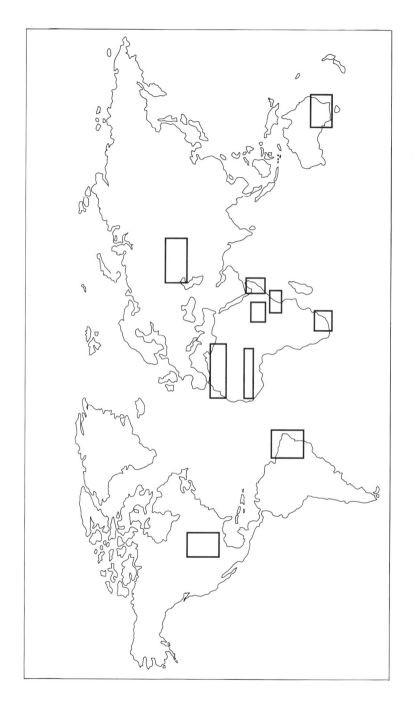

Figure 1. Location of case studies.

PART I
Drought, desertification and food production

Drought, desertification and food production

MICHAEL H. GLANTZ

Drought

Throughout history, most human settlements, however small or prosperous, have had to contend with drought. A few thousand years ago people wrote about how adverse weather conditions affected food production and water supplies. The idea of a genesis strategy was suggested in biblical times: to store surplus grain produced in the good rainfall years for use in the years of poor rainfall. This idea reappears during prolonged severe drought episodes to remind societies that droughts recur and that they must learn from history, or they will repeat their mistakes.

The twentieth century has seen its share of droughts. One could argue that the early 1970s were a turning point in global awareness about the need to understand better the drought phenomenon, its causes and consequences, and to develop mitigation strategies to cope with its consequences. More specifically, it was prolonged droughts in the West African Sahel and other locations around the globe in the early 1970s that served to inspire the publication *The Genesis Strategy*,[1] a most recent reminder of an ancient drought-coping mechanism. The occurrence of severe droughts throughout Africa and in India, North America, China, the Soviet Union, Australia, and western Europe in the 1980s once again underscored the vulnerability of developed and developing societies to drought. Drought has been the subject of a great deal of systematic study, particularly reconstructions of drought history, computations of drought frequency, and, to a lesser extent, investigations of first-, second-, and even third-order impacts of drought on society.

Even in today's age of high technology and instant communication, agricultural and livestock production in industrialized as well as developing societies can be sharply reduced by drought-related stresses. People show little surprise when drought-related crop failures occur in Third World countries. However, industrialized countries, such as the United States, Canada, and Australia, have also been unable to drought-proof their agricultural areas, in spite of attempts to do so. Thus, no country can claim to be immune to the uncertainties of seasonal or annual rainfall. Although we have little capability (if any) to avert meteorological drought, reliable and credible information about drought and its impacts could be used by individual farmers as well as national-level planners and political leaders to improve society's ability to minimize the scope and severity of its consequences.

The most popular perception of a drought is as a 'meteorological' phenomenon, characterized by a lack of rainfall compared with an expected amount for a given period of time. For some, a drought exists when rainfall is 75% of the long-term average for a given period; others might consider it to occur at 60% or 50% of 'normal.' In reality, definitions of what constitutes a drought will need to vary, depending on one's expectations about moisture needs for specific human activities within a given area. Thus, there can be agricultural drought and hydrological drought, as well as meteorological drought, with different people using the same information about precipitation disagreeing about whether a drought has in fact occurred.[2] While it is not important to have agreement on a universal definition, it is important that people know what others mean by 'drought.'

Some researchers have suggested that drought should be defined in terms of societal factors. They have argued that, before a reduction in precipitation becomes a concern to society, it usually adversely affects the established economy of the region.[3]

For example, the spring wheat region of the Canadian prairie provinces (Alberta, Saskatchewan, and Manitoba) was plagued in 1974 by several adverse weather conditions throughout its growing season, including a late wet spring planting and an early wet harvest period. Another adverse condition that year was a prolonged midsummer drought. Local newspapers were filled with pleas from distraught farmers requesting drought assistance from their government. Post-season statistics later showed that wheat production that year had in fact dropped by about 20%, supporting farmers' fears that drought had adversely affected their agricultural productivity. However, by the end of 1974, grain prices in the international marketplace had risen to their highest levels ever. Many of those complaining farmers who were ultimately able to produce a wheat crop made more money from their harvest than they had been able to generate from their crops in 'normal' rainfall years.

A few years later, drought once again occurred in the Canadian prairie provinces. Interestingly, a list of previous devastating drought years in Canada failed to include the 1974 situation, despite the fact that it could be classified as a meteorological and an agricultural drought. It was no longer treated as a drought year, because it did not have an adverse impact on the regional or national economy.

Droughts often become highly visible when they are associated, rightly or wrongly, with famine. The truth is that governments prefer to blame natural factors such as droughts for extreme food shortages. If nature is to blame, then governments need not bear the direct responsibility for having been unable to supply their citizens with adequate nutrition. Yet, for the most part, droughts can occur without precipitating a famine situation; and the historical record has clearly shown that famines have frequently taken place in the absence of drought conditions.[4] An important point that must be made time and again is that droughts by themselves seldom lead to famine. Often, drought, a 'creeping' phenomenon, combines with other underlying societal and environmental conditions to produce famine-like conditions. In 1992 famine became clearly disconnected from drought as the principal causative agent. The reason 1992 is cited as the critical year for understanding famine is threefold: Bosnians, Somalis, and Kurds. Here one finds starving people in three situations in three different parts of the world (Europe, Africa, and the Middle East, respectively). In each case, weather has played little or no role, while military conflict played a major one. Though governments would prefer to have blamed their inability to feed their people on the weather-related problems, these three 1992 famine situations do not allow them to get away with it.

Drought has also been blamed for prompting mass migrations, environmental degradation (often referred to as desertification), and internal unrest. While drought may have been an important factor in each of these processes, it often proves to have been but one of many intervening factors. Bradford Morse, former Administrator of the United Nations Development Programme (UNDP), suggested in the mid-1980s that 'drought itself is not the fundamental problem in sub-Saharan Africa ... The present drought, however, intensified the interaction of the factors impeding development in Africa; it has laid bare the African development crisis.'[5]

An international report on famine went a step further, noting that 'At each stage from its genesis in rural poverty and food-production failures through to the reduction of communities to destitution and starvation, famine is avoidable. More than that, its causes are much more complex than just bad luck with the weather.'[6]

This view is reinforced by longitudinal studies that have shown that the impacts of drought can differ greatly, even when they take place in the same location but at different times. The US Great Plains would serve as such an example, as would the West African Sahel.[7] Studies also show that different socioeconomic groups within the same area are affected by drought in different ways, underscoring the fact that not all people are equally affected by adverse weather or by drought-related famine situations.[8] Discussion of the inequitable and varying impacts of climate variability and climate change was the focus of a recent United Nations Environment Programme Workshop, 'On Assessing Winners and Losers in the Context of Global Warming.'[9]

Desertification

The word 'desertification' was apparently first mentioned in a 1949 report on 'Climate, Forests, and Desertification in Tropical Africa' by Henri Aubreville, a French forester. Since then, it has taken on scores of definitions, which can be divided into two categories: those that view it as a process, and those that view it as an end state. Desertification, like drought or famine or, for that matter, other 'creeping' phenomena, can be viewed as a process of degradation of land resources which can manifest itself in a variety of ways such as declining crop yields, decreasing density of vegetative cover, reduced soil fertility, and so forth. It can also be viewed as the creation of an unproductive desert-like landscape in a place where none had existed in the recent past.[10] The latter image of desertification prevails among the general public.

Desertification as a concept, popularized by the environmental degradation witnessed during the Sahelian drought in West Africa in the early 1970s, is often linked to droughts. However, studies have shown that desertification sub-processes (e.g., wind and water erosion) frequently occur in the absence of drought. While droughts can exacerbate and even highlight desertification sub-processes at the regional and local levels, they are not necessarily a precondition for the initiation of those processes.

Issues of environmental and societal vulnerability to drought are frequently related to the way in which marginal lands are used by society. For example, land-use practices that are effective in humid regions are often not as effective in arid and semiarid areas. Nevertheless, emigrants from the wetter areas often rely on practices that seemed to work under wetter conditions but had not been tested in marginal areas. The end result of this mismatch between land-use practices and the long-term rainfall characteristics (or other features that make a region marginal for sustained agricultural production) is environmental degradation and desertification.

Drought and desertification can occur separately, without one necessarily causing or being caused by the other. One important determining factor with regard to whether these have a relatively high potential for linkage is the condition of the land and its vegetative cover at the time of drought. In this regard, Monique Mainguet observed that 'it seems that drought is more the revealer than the cause of land degradation.'[11] Land degradation resulting from inappropriate land use can render the soil more vulnerable even to the 'normal' variability of climate from one year to the next. For example, crop residue left in the ground as stubble following a harvest makes soils less vulnerable to wind and water erosion. In the absence of protective measures in areas objectively determined to be susceptible to desertification, however, drought can initiate or intensify desertification processes. Thus, the loss of soil fertility, reduction of fallow time, soil trampling by livestock, deforestation, and tree-cutting, each of which lowers soil moisture, can heighten the adverse impacts of meteorological drought and increase the likelihood of agricultural drought. As Baker noted, 'an environment weakened by overuse will be much more prone to collapse during normal stress.'[12] Thus, agricultural activities in marginal areas can increase the risk of environmental degradation as well as crop failures during prolonged periods of moisture stress.

Scientists have developed hypotheses about how rain-producing atmospheric processes are weakened by the reduction of vegetative cover in arid and semiarid areas because of increases in albedo (reflectivity of the Earth's surface). Jule Charney and colleagues, in a search for causes of the lengthy drought in the West African Sahel in the early 1970s, described the feedback processes between the land, vegetation, and the atmosphere as follows:

> the radiative heat loss caused by the high albedo of a desert contributes significantly to the sinking and drying of the air aloft and therefore to the reduction of precipitation. This dependence of precipitation on albedo led [Charney] to propose a biogeophysical feedback mechanism linking vegetation, albedo and precipitation as a partial explanation for recurrent drought in areas bordering deserts. If the soil is light, dry and sandy, as it often is in these areas, a decrease in vegetation will lead to an increase of albedo, a reduction of precipitation, and therefore a further decrease in vegetation or at least a perpetuation of the initial decrease.[13]

Hypotheses like these tend to reinforce a scientific as well as a popular belief that desertification processes in regions like the Sahel can contribute to the perpetuation of regional drought conditions.[14]

In the past few years challenges to the original albedo effect hypothesis have appeared, based on evidence that direct measurements in the Sahel have not indicated an increase in albedo, as hypothesized by Charney and co-workers.[15] More recently, Tucker and his colleagues[16] have suggested that the desertification process in the West African Sahel may actually have reversed in the 1980s. However, recent research findings have again come out in support of Charney's hypothesis.[17] In addition, a recent research paper suggested that climate changes at the regional and local levels associated with desertification processes may confound the signal for a greenhouse-gases-induced warming of the lower atmosphere, popularly known as the greenhouse effect. This confounding factor, adversely affecting air temperature measurements (along with other such factors such as the urban heat island effect and changes in the location of meteorological stations), will also prompt people to challenge the temperature readings in desertified areas.[18] The jury is still out on this potentially important scientific issue.

Food production: the role of marginal lands

Aside from either the exploitation or development of non-conventional food sources, there are two basic ways to increase agricultural food production: (1) to increase crop yields, and (2) to expand the hectarage under cultivation.[19] These two basic strategies, however, will not necessarily resolve a nation's food production problems, as will be shown in the subsequent case histories. In fact, their applications have frequently exacerbated the plight of cultivators, pastoralists, and their governments in the long term by leading to unintended environmental changes that work against the prospects for sustainable increases in food production.

Increasing crop yields can be achieved by irrigating previously unirrigated fertile lands, by relying on improved technologies or agricultural production methods such as fertilizer and pesticide application to improve yields on existing fields, or by developing techniques to better match the appropriate crops with existing rainfall regimes. The development of high-yield varieties (HYVs) of grain, bringing about yield increases several times that of ordinary varieties, was seen in the 1950s and 1960s as the beginning of a

'Green Revolution.' Optimism was high among government leaders, agricultural development specialists, and farmers, who believed the new varieties would resolve once and for all chronic regional food shortages throughout the Third World.[20] The productivity of HYVs in India and Mexico compared favorably with that on farmlands in developed countries. It soon became apparent, however, that the Green Revolution created a whole new set of technological, economic, and social problems; for example, with HYVs there was a need for costly inputs such as water, irrigation facilities, fertilizers, herbicides, and pesticides, all of which only wealthy farmers could afford.

Because such inputs would be too costly for the production of subsistence food crops in sub-Saharan Africa, governments and foreign assistance agencies chose to support development projects that were based on the production of HYVs of grains new to the region or of cash crops (such as sugar cane, coffee, and cotton). Dr Mustafa Tolba, former Executive Director of the United Nations Environment Programme, noted some years ago that traditional African crops such as sorghum, millet, chickpea, pigeonpea, and groundnut had generally been neglected by development experts.

> Because these five crops are largely grown for food rather than for cash, national and international agricultural programmes have paid them scant attention ... The poor farmers who grow these crops cannot afford the fertilizers or irrigation which their wealthier counterparts use to produce cash crops.[21]

One result of the Green Revolution has been the takeover of productive land by wealthier farmers at the expense of poorer ones, who as a result have been forced to migrate to urban centers in search of work or to increasingly marginal areas in search of land. HYVs and cash crop production also encourage the use (as well as destruction) of marginal lands:

> In some countries, an increasingly critical balance of payments situation has forced governments to plant more and more cash crops. If no more arable land is available, the new

crops have to be planted in marginal land, previously pasture, which is brought under the plough for the first time. Much of the thin soil may be lost in the first heavy rain which occurs when the field surface is still bare.[22]

Even if one were to stamp a 'success' label on the Green Revolution on the Indian subcontinent, researchers have questioned attempts to re-create the same revolution on the African continent. A Green Revolution has not taken place in many parts of sub-Saharan Africa. Dr Edouard Saouma, Director-General of the United Nations Food and Agriculture Organization (FAO), noted that

> neither the green revolution technology of Asia nor the capital-intensive methods of western agriculture have proved viable in the very different conditions prevailing in most of tropical Africa. Soils are generally more fragile and more easily eroded than in other regions.[23]

The successes of the HYVs, such as they have been, may prove to be region-specific, with the experiences gained in one region not necessarily being directly and unquestionably transferable to other regions without considerable modification.[24]

Reports have suggested that increases in crop yields witnessed in the 1960s and 1970s have begun to taper off.[25] In the absence of a second Green Revolution or a reduction in the cost of required inputs or easier access to credit by poor farmers, pressures will mount on cultivators to bring presently uncultivated marginal lands into production. While this may be the least expensive way in the short term to increase food production in developing regions within countries, it greatly increases the level of risk of land degradation and of crop failures in the long term.

The linkages between the HYV option and the option of expanding the hectarage under cultivation are becoming increasingly more obvious, as was noted in an FAO report:

> Since new arable land in the developing countries will be getting more scarce – by 2000 almost two-thirds of the developing-country population are projected to live in countries with

95% or more of their total potential arable land under the plough – higher yields are the answer for most countries.[26]

The remaining areas available to meet the increasing demand for cropland are those that in the past have been considered too marginal for agricultural production.

It is important to note that in most countries with arid areas there is an imaginary line, often approximated by mean rainfall isohyets (lines of equal precipitation), that separates the areas with a rainfall regime capable of supporting sustained rain-fed cultivation from those that are mainly suitable for livestock grazing. In the West African Sahel, for example, the 400–500 mm isohyets have been suggested as such a border. Some authors have referred to that boundary as a 'line of tension' between cultivators and pastoralists, with the cultivators dominating the regions with more than 400–500 mm of rainfall per year (on the average) and the pastoralists dominating the region toward the Sahara with less than 400–500 mm. Conflicts have occurred (and continue to occur) between cultivators and pastoralists, as one group encroaches on the land of the other.

High population growth rates also add to pressures on the land, as do conflicts between groups competing for control and use of scarce land and water resources. As population increases among cultivators (at higher rates than among neighboring pastoralists), there is pressure to shift that 'line of tension' toward the lower rainfall areas. Cultivators, often with the encouragement of their own governments, move onto the rangelands to put additional land under the plow. This in turn causes pastoralists to move (further north, in the case of the Sahel) into the increasingly drier interior of the continent. Both population shifts have increased the degree of land abuse which becomes highly visible during periods of drought.

> In many developing countries, an expanding population requires more food and has increased pressure to expand the cultivated area. This pressure, coupled with political expediency and lack of scientific expertise, has resulted in more ... utilization of marginal land ... Thus, the expansion of agriculture on steep hillsides has led to serious erosion in Indonesia

and Kenya; increasing pressure of slash-and-burn agriculture is destroying tropical forests in the Philippines . . . and overgrazing and deforestation is contributing to the southward march of the Sahara.[27]

Another pressure to move into agriculturally marginal areas results when cultivated land can no longer support the increasing population.

> Availability of arable land, until relatively recently, was sufficient in most countries to sustain the traditional (land extensive, low input, rotational, long fallow period) agricultural systems without presenting major problems. Today, however, increased population pressure has resulted in increasing pressure on the land. The need for increased production has led to expanded use of marginal land with low and unreliable productivity. In addition, fallow periods have been reduced leading to even further declines in yields. Savannas, which traditionally have been used for herding, are now being converted to permanent cultivation. These factors contribute to serious degradation of the natural resources.[28]

However, an increase in population is not always a necessary condition for expansion of agricultural activities into marginal areas. Many other factors are at work that tend either to 'push' or to 'pull' cultivators into such areas.

Recall that 40 years ago, Nikita Khrushchev, then First Party Secretary of the Communist Party of the Soviet Union (CPSU), launched his Virgin Lands Scheme in an attempt to increase sharply his country's grain production. The scheme's success would have demonstrated to the developing world that the Soviet Union was as much an agricultural force as an industrial power. The plan 'encouraged' people to move into western Siberia and northern Kazakhstan to put large tracts of semiarid land under mechanized agricultural production. Soon stories of the difficulties, if not failure, of this approach reached the press, as people began abandoning the virgin land areas to leave the inhospitable environment. After a few years of favorable harvests, drought conditions plagued the virgin lands areas, and since that time rain-fed agricultural activities have been supplanted by more appropriate farming practices. The architects of the virgin lands strategy failed to take seriously the marginality of the climate for sustained agricultural production; they paid for it at the time, with humiliation in the economic development community. In fact, Leonid Brezhnev, later to become the General Secretary of the CPSU, was directly involved in the scheme's implementation. A book about the problems encountered in the virgin lands area was prepared under his name by a ghost writer.

Thus, farmers also move into the agricultural 'margins' as a result of socioeconomic or political pressures, such as government policies to increase food production or to develop cash crop and irrigation schemes. Such policies often require the takeover of the relatively more productive lands, displacing subsistence farmers. Once displaced, these people must find new lands on which to grow food. For example, although there have been several major droughts in the West African Sahel this century, 'the difference in recent years has been that, with the best land devoted to cash crops, ordinary farmers are driven onto marginal land, which is most prone to soil erosion and "desertification." The result is a vicious circle of degradation that serves only to intensify the drought.'[29]

Moving into marginal areas is often a preferred option by farmers (and pastoralists) who are reluctant to migrate from rural to urban areas, choosing instead to go from one rural area to another. Other pressures that cause farmers to move into increasingly marginal areas relate to declining soil fertility resulting from widespread soil degradation on rain-fed croplands in the relatively more productive areas. In addition, there are several examples of formerly productive irrigation schemes that have become unable to sustain agriculture (or yields) over time as a result of salinization, alkalinization, or waterlogging, and have been abandoned. These pressures, among others, result in a 'push' of cultivators (and, in turn, of herders) into increasingly marginal areas.

There are also factors that attract (i.e., pull) people out to the margins. As a result of a favorable wet period, for example, governments tend to encourage the exploitation of these lands and often overlook or misjudge the long-term difficulties in sustaining agricultural production there. Farmers and herders during wet years are also lulled into a false sense of security based in large measure on wishful thinking. Yet another factor that attracts people to the marginal lands is the belief that technology (e.g., irrigation or deep-well construction) will provide society with a drought-mitigating or drought-proofing capability that had not previously been available. Mechanized farming is yet another technological factor that pulls people toward the marginal areas. In theory, with mechanization large expanses of land can be put into cultivation. However, citing an example from the Sudan, Schulz noted some dangers:

> tractor-driven ploughs can quickly remove topsoils from fragile lands. Also, farmers who have machinery can actively farm larger areas, contributing to land degradation by (1) extending their operations into fragile or semiarid land areas and (2) by pushing herdsmen from stronger grazing areas into more fragile ones.[30]

Cultivating the dry margins: drought follows the plow

The recent population movements into marginal areas have interesting historical precedents. The Great Plains region of the United States (Figure 2) is an example of a dryland margin. During the westward migration of settlers across the North American continent during the nineteenth century, this treeless grassland was put under the plow with consequences that were remarkably similar to those witnessed today in many arid and semiarid parts of the world.

During the early part of the nineteenth century, explorers and prospective settlers coming from the humid, eastern part of North America identified the Great Plains as the Great American Desert. One explorer after another perceived that there was little in the region of agricultural value for potential cultivators. Their journals and reports underscored their perceptions of the apparent uselessness of the Great Plains for agriculture. The region was seen as hot, dry, and windy with poor soils and water resources inadequate for sustained cultivation.

However, after the early 1860s that dominant perception, generated in the first half of the nineteenth century, began to change

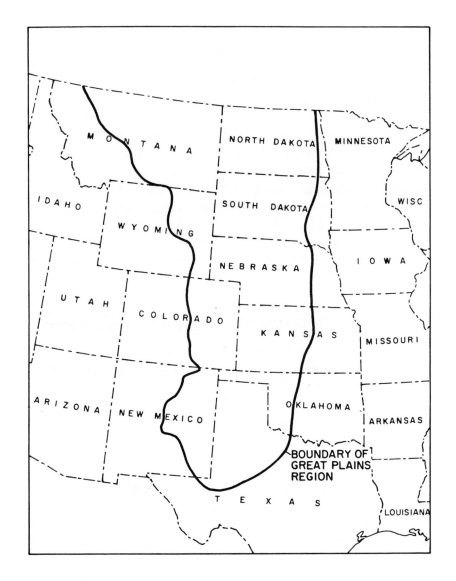

Figure 2. The Great Plains region.

dramatically as rains returned to the Plains. As a result of the newly prevailing climatological condition, many of the pioneers who were moving westward across treeless grasslands chose to settle there as farmers. Waves of new migrants were attracted to the region following the return of rainfall amounts favorable for agricultural activities.[31] Settlers brought with them technologies and techniques suited to the humid areas from which they emigrated; these, in time, proved unsuitable to the arid and semiarid conditions of the plains.

In the 1870s and 1880s there was widespread discussion in North America about the positive influences of trees on rainfall. Scientists and farmers alike believed that rainfall in the Great Plains had been increasing as a result of tree-planting, itself the result of human settlement. It was then that the notion of 'rain follows the plow' became popular.[32] This notion attracted an increasing number of easterners into the region in search of a chance to improve their quality of life or in search of change and adventure. The 'rain follows the plow' notion also sparked legislation by the US Federal Government fostering tree-planting throughout the region. Before long, the belief was expanded to include the notions that settlements and cultivation also influenced atmospheric processes to produce increasing amounts of precipitation across the Great Plains. Partly as a result of the combination of several years of good rains and the agricultural successes of early settlers, a belief emerged that rain had indeed followed the plow. In the words of American historian W.P. Webb, 'in the Great Plains proper there was as yet little thought of general irrigation. The people were settling there under the illusion that rainfall would follow agriculture.'[33]

Webb summarized the prevailing belief about the relationship between rain and the plow, when he wrote that 'the plowing up of the land would hold the moisture, would increase evaporation, and would make precipitation possible. The growing of crops would in some way have the same effect. The burning of the prairie would produce rain.'[34] According to Webb, this belief stemmed from two basic factors: 'the first is that men hope that rainfall will increase, and this hope is father to the thought that it has increased; the second is that we have precipitation cycles which last over a number of years.'[35]

Soon, however, the notion fell into question, as parts of the Great Plains experienced severe droughts in the last few decades of the nineteenth century. These severe droughts prompted settlers to abandon their homesteads and everyone to reassess the belief that rain follows the plow (Figure 3). 'The undependable rainfall posed a problem that for two decades and more proved insoluble. Time and again, between 1870 and 1890, settlement advanced far out upon the plains in periods of relatively high rainfall, only to be forced back by the dry period which always followed.'[36] More specifically, 'half the population moved out of western Kansas between 1888 and 1892, with heavy exoduses from comparable parts of Nebraska and the Dakotas.'[37] Part of the problem generating crop failures in the region was that 'in the nineteenth century, crops grown by methods employed in humid regions were usually not successful when yearly precipitation on the Plains fell below twenty inches, and they failed in times of drought.'[38]

The worst American drought in the twentieth century occurred in the 1930s, when large rainfall shortages for several years combined with economic depression to cause widespread migration out of various parts of the Great Plains. Large dust storms accompanied these droughts, causing the southern part of the US Great Plains to become popularly referred to as 'the Dust Bowl.'[39] The plight of farmers in the Dust Bowl was recorded by many photographers, artists, and writers of the time and was especially well captured by John Steinbeck in his novel, *The Grapes of Wrath*.

The specter of widespread destruction of productive farmland in the Great Plains precipitated a flurry of government-sponsored reports to identify the causes and to develop measures to protect future generations from the impact of such droughts.[40] Some proposed measures included the development of appropriate land-use classifications (i.e., matching agricultural and herding practices to the prevailing climate and soil conditions), improved tillage practices, the planting of shelterbelts of trees, and the encouragement of strip farming to arrest wind erosion and soil desiccation caused in part by the ever-present hot, dry winds during the growing season.[41]

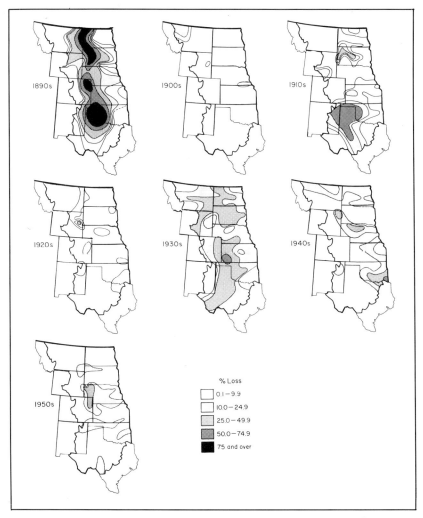

Figure 3. Shifts in the population of the Great Plains from the 1890s to the 1950s. Reprinted from Bowden *et al.*, 1981. In: T.M.L. Wigley, M.J. Ingram & G. Farmer (eds.), *Climate and History*. Cambridge: Cambridge University Press.

With a return of favorable weather conditions in the 1940s, along with an increase in grain prices and an increase in the use of groundwater for irrigation, agricultural production in the region increased. However, a recurrence of drought in the mid-1950s, while not as devastating as that of the 1930s, reminded people that regional land resources were fragile, especially in the southern Great Plains centering on the Texas Panhandle. Yet another flurry of government reports warned farmers against inappropriate land-use practices like those that had exacerbated the impacts of the 1930s drought. The 1950s drought prompted even more farmers to shift to irrigation, further developing a dependence on seemingly abundant groundwater from the Ogallala Aquifer (Figure 4).[42]

By the late 1960s, high levels of agricultural production, largely as a result of technology (especially irrigation) and of extremely favorable climatic conditions, had caused a glut of grains in the international marketplace that in turn led to a marked decline in grain prices. Farmers in the United States were paid by the federal government to take land out of production – that is, they were paid not to grow certain crops already in surplus. A similar situation occurred in the Canadian portion of the North American Great Plains. A government plan – LIFT (Lower Inventory for Tomorrow) – was proposed to remove cropland from production in the prairie provinces at the end of the 1960s. For a few years it appeared that global food production problems were soon to be solved.

In 1972, a year notable for its worldwide climate anomalies, the global food situation changed abruptly and for the worse. While the United States did not suffer from drought, US farmers, encouraged by their federal government, responded to the worldwide increase in the demand for grains by plowing up agriculturally marginal lands to produce grain crops (especially wheat), taking advantage of the highest prices ever for that commodity. Unfortunately, Great Plains farmers neglected lessons they had painfully learned during the 'Dust Bowl' days, as shelterbelts and hedges surrounding cultivated fields were torn down and strip farming was abandoned. In addition, land that was more suitable for grazing activities than cultivation, was again put under the plow. Meteorological and agricultural drought and even major dust storms reminiscent of those of the 1930s returned to the region in the mid-1970s (Figures 5 and 6). The dust storms, visible from space, generated concern that farmers had once again resorted to ecologically destructive agricultural practices.[43] An important question

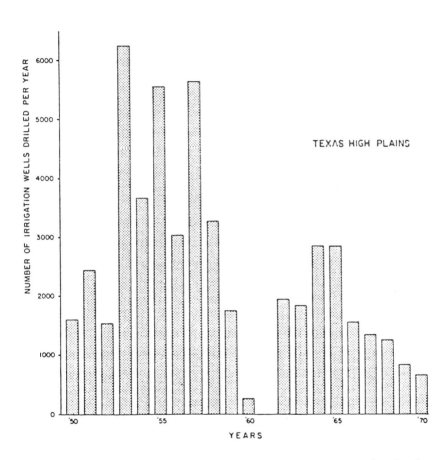

Figure 4. Irrigation wells drilled in the High Plains of Texas and Colorado, 1950–70.

Figure 5. Photograph from GOES-1 geosynchronous satellite, February 23, 1977, showing thick dust raised by strong winds over eastern Colorado, western Kansas and Oklahoma, and west Texas. The dust boundary is nearly coincident with the Texas–New Mexico border.

that should have been raised is not *what* should have been planted or even *how* or *when* the planting should have been done, but *whether* it should have taken place at all in those parts of the Great Plains known to be marginal for sustained crop production.

The drying out of the US Great Plains at a level reminiscent of the situation during the drought of the 1930s has been proposed as a possible regional scenario that could result from an increase in global temperatures because of an increase in atmospheric carbon

Figure 6. A piece of abandoned farm equipment in a field in eastern Colorado has trapped top soil eroded by wind. The photograph was taken in March 1977, following major dust storms reminiscent of those of the 1930s Dust Bowl.

dioxide.[44] As a result of this possible analogue to global warming, the 1930s period captured the attention of several researchers who in the 1980s and 1990s sought to gain a glimpse of the possible consequences for the 'American breadbasket' of a warmer atmosphere. The Dust Bowl days have been used as a basis for generating credible (but not necessarily reliable) scenarios about the possible impacts of global warming on the region's economic activities, especially agricultural production.[45]

Whenever drought returns to the Great Plains, one is reminded not only of the 1930s drought and the consequences of inappropriate land use but also of the nineteenth-century debate about whether large segments of that region in the long run would revert to an unproductive desert landscape (as the arid and semiarid US

West had originally been perceived) or would be a sustainable garden (as the humid eastern US had been perceived). The trend toward depletion of groundwater in the southern reaches of the Ogallala Aquifer, which underlies much of the Great Plains, tends to heighten concern about the region's long-term fate. Depletion of this aquifer in its southern extremities forced some farmers by the early 1980s to revert to dryland farming, to change the mix of crops they grow and how they water them; some farmers even abandoned their farms. Two authors (with little support from any quarter) have recently proposed a government-supported abandonment of large tracts of farmland in the Great Plains in order to let the land return to its earlier natural state – complete with grasslands, roaming buffaloes and all.[46]

In recent years similar concerns have been raised about the adverse impacts of demographic pressures on arid and semiarid regions around the world. In Part II the situations in the West African Sahel, Somalia, the Brazilian Northeast, Kenya, Australia, Ethiopia, Northwest Africa, Virgin lands of the former Soviet Union, and South Africa are discussed. The movement of people into agriculturally marginal areas is a reality shared by both the rich and the poor nations of the world. The consequences, however, are considerably more devastating in human terms in the developing countries than in the industrialized ones, owing to the latter's economic capability to date to 'buy' their way out of their ecological problems.

PART II

Case studies and conclusions

The West African Sahel

MICHAEL H. GLANTZ

'Sahel' in Arabic means coast or shore. It is used in this region because the Sahel straddles the southern edge of the Sahara Desert, the largest desert in the world. Some have defined the Sahelian zone by the types of natural vegetation found in the region. Others have defined it in terms of rainfall isohyets or lines of equal precipitation. To a large extent there is a correlation between rainfall regimes and types and amounts of natural vegetation, depending on the types of soils in a given area.

The Sahel is an arid/semiarid area that can generally be defined as the area encompassed approximately by the 250 and 750 mm isohyets (Figure 7). In such areas with relatively low average annual rainfall amounts, rainfall within a year and between years is highly variable. The consequences of such variability ripple through ecosystems and societies, and especially affect attempts at rain-fed agricultural production. The Sahelian growing season usually begins in June and ends in September, but lengthens at both ends of the season as one moves away from the Sahara southward toward the West African coastline. The risk of crop failure because of lack of rainfall (a meteorological drought) also diminishes as one moves southward. The amount of rainfall in the region sharply decreases as one moves from south to north, meaning that even small perturbations in rainfall patterns can have major impacts on agricultural production and livestock rearing.

The main activities of people in the Sahel are rain-fed farming, pastoralism (nomadic people who move with their herds), and

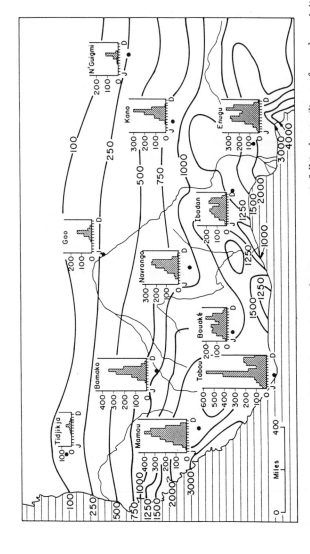

Figure 7. Climatic zones of the Sahel–Sudan region, indicating average rainfall isohyets (lines of equal precipitation).

agropastoralism (farmers who also keep livestock). Various forms of irrigation are resorted to in the region, ranging from capital-intensive, highly developed schemes with sophisticated irrigation technology that produce cash crops for export, to labor-intensive subsistence farming along the banks of rivers as they recede in summertime.

Aside from scholars and economic development specialists, few showed any special interest in the West African Sahel until the early 1970s, when the region was catapulted into prominence by the international news media. Food shortages and, later, famines had plagued the region as the result of a 5-year drought that began in 1968, with the last year (1973) being the most severe. The plight of the region's inhabitants was exposed not by governments, United Nations (UN) agencies, or private voluntary organizations, but by French newspaper reporters.[1] West African governments were reluctant (if not embarrassed) to let the world know that they could not feed their citizens. Foreign governments that were potential aid donors had also been reluctant to go against the UN rule of 'non-interference in the domestic affairs of independent countries' and for one reason or another chose not to pay attention to the consequences of the emerging food crises. Once the famine conditions were exposed, horror stories filled the news media of thousands of human and livestock deaths and tens of thousands of emaciated peasants flocking to refugee camps or to urban centers, begging for food. Photographs of once-productive areas that had been transformed into denuded, desert-like landscapes accompanied those news stories.

At first the disaster was considered a natural one: meteorological drought had returned to the region (at the time it was the region's third major drought in the twentieth century). But closer scrutiny of the facts showed that human activities (including policies) had unwittingly made the bad consequences of a drought situation even worse.[2] Clearly, the mix of a natural phenomenon (drought) with inappropriate land-use activities in marginal areas had led to widespread environmental degradation and desertification throughout the West African Sahel, where rain-fed agricultural practices encompass about 96% of the region's cultivated area. It was estimated that, by the mid-1970s, 100 000 to 200 000 people had perished along with about 12 million (or half) of the cattle in the region.[3] In addition, large expanses of marginal lands were considered to have been ruined for future use by either farmers or herders because of severe soil erosion, a major form of desertification.

Assessments of the climatology of the West African Sahel have clearly shown that the 15-year period preceding the onset of drought in 1968 was one with favorable rains (above the long-term average).[4] Following such wet periods, a return to below-average rainfall conditions was to be expected – a general feature of rainfall in arid and semiarid lands.

Some debates developed in the scientific community over (1) whether the 1968–73 drought had actually ended in 1973, (2) whether the region's climate was changing (becoming more arid), and (3) whether the primary causes of famine and environmental destruction were natural or human-induced.

(1) On the rainfall issue, many scientists have contended that the drought which began in the late 1960s continued well into the 1980s. Figure 8 depicting a regional rainfall index makes their point. However, on the ground, Sahelian farmers believed that they had a reprieve from meteorological drought for a few years beginning in 1974. Food production improved for a few seasons, followed by a return to drought in 1977 lasting until the mid-1980s.

(2) Some researchers argued that a 'change' in West Africa's climate had taken place.[5] Derek Winstanley, for example, suggested that the continuous drought had lasted for 17 years and the chance of such a prolonged drought occurring without a climate change was extremely low. However, statistical analysis of the region's rainfall record suggests that the probability of a drought of such length occurring in the Sahel had not necessarily changed; even without a change in average rainfall conditions, lengthy drought episodes could still occur.[6]

Today, there are differing views on drought in this region (and in other parts of Africa as well). Are they a part of the 'normal' climate regime or apart from it? Are the droughts

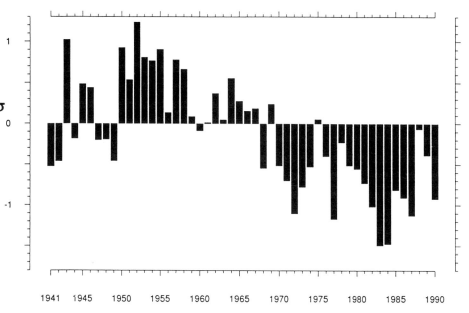

Figure 8. Time series plot of yearly rainfall index values for the Sahel, 1941–90. Reprinted with permission, *Journal of Climate*, 1993, **6**, 2196.

induced by human activities? Are they decadal-scale fluctuations, or the result of the regional consequences of a global warming of the atmosphere?

Some observers suggest that, with a climate change, areas in Africa that are presently dry will get drier while those that are wet will probably get wetter. Others have suggested that climatic zones will shift in such a way that the presently wet areas may become drier and dry areas will stay dry. Still others believe that wet areas will get drier and dry areas will get wetter.[7] Educated guesses notwithstanding, the forecasts for possible future changes in African regional climates are still quite unreliable. In the meantime, while physical scientists focus their research on natural factors in order to improve their understanding of the regional consequences of a possible global climate change, demographic changes with adverse impacts on the environment tend to confound attempts at early detection of regional climate change.

(3) In the mid-1980s severe food shortages plagued about 27 countries in sub-Saharan Africa, including those in the West African Sahel. Famines, however, appeared only in Chad, Sudan, Ethiopia, Mozambique and Angola. At that time these were also the only governments facing internal wars.

During the droughts of the mid-1980s, the West African Sahel had been spared from a famine situation most probably by the fact that Western aid agencies had become established in the region following the famine in the early 1970s. Today, famine in the region is less likely, even if the region is faced with a drought of the same intensity and magnitude as that of the early 1970s, because of new political arrangements and coordination between donors and recipients and because of early warning systems established in the mid-1980s. As Rolando Garcia noted,

> the effect of droughts with similar natural characteristics may therefore be shown to vary historically as a function of the dynamics of the production system and the corresponding frame of social relations. *It is therefore clear that purely climatological parameters do not suffice to characterize a drought situation.*[8] [Italics added]

During the wet decades of the 1950s and 1960s, government planners encouraged farmers to move northward toward the newly appearing green areas that were situated along what had been the arid edges of the Sahara. The areas into which cultivators moved were increasingly marginal for long-term sustained cultivation the further north they moved.[9] The soils and long-term rainfall conditions of these grasslands were more suited to livestock grazing activities than to dryland farming.[10] Pastoralists in turn were displaced from their traditional rangelands, moving further northward.

At the time there seemed to be few, if any, environmental problems associated with these migrations. Rains were at least temporarily favorable to agriculture and rangeland productivity. Because of the rains, many settlers were lulled by nature and by their governments into believing that these areas could sustain agricultural activities well into the future. The reality of the situation, however, was that these newly occupied areas were subject to alternating runs of wet and dry years. Although these fragile areas could sustain human activities periodically (for short periods), those activities could not be sustained in the long term in the absence of human intervention such as irrigation. An added negative factor was that the consequences of inappropriate land-use activities during extended wet periods were setting up farming and herding activities as well as local ecosystems for eventual collapse with a return to below-average rainfall.

For example, the relocation of a nomadic population in the Sahel (the Tuaregs) to lands more susceptible to the effects of reduced precipitation further increased their vulnerability to meteorological drought.[11] More generally, David Campbell noted that, 'the expansion of the cultivated area was a response to increased population, to the sedentarization of nomadic groups and was facilitated by successive years of favorable rainfall in the 1960s.'[12] As a result of this expansion, the nomads became trapped on temporarily improved but still extremely marginal rangelands, 'sandwiched' between the Sahara Desert to the north and the encroaching cultivators to the south (Figure 9). During runs of favorable rainfall years, pastoralists were able to exploit even these very marginal regions with relatively large herds. During prolonged dry periods, however, the impacts of inappropriate land use became exposed. In other words, during wet periods farmers and herders were able to get away with improper land use that could not be tolerated by the terrestrial environment during dry spells and droughts.

Because water resources had been (and still are) seen as *the* limiting factor for human settlements as well as for sustainable herd size in dry years, hydrological solutions were sought to resolve what had been perceived as hydrological problems. A major response by governments and development agencies before as well as after political independence of Sahelian countries was to provide deep wells throughout the region. Deep wells would ensure adequate water supplies at the local level. While the wells were constructed to provide year-round water supplies, which they successfully did, little consideration was given to the imbalance that would develop between new water supplies and existing vegetation during drought episodes. Reports in the early 1970s from the Swedish International Development Agency suggested that many

Figure 9. Degraded grazing land in the Sahel, where nomads became 'trapped' between the encroaching Sahara Desert and the northward-moving cultivated areas to the south.

of the livestock deaths recorded during droughts in the West African Sahel resulted primarily from hunger (i.e., lack of vegetation), and not from thirst (i.e., lack of water). What government officials and water experts had failed to consider was that the existing balance between traditional technology, water resources, and vegetation would be destroyed by importing new advanced water-related technologies, making factors other than the availability of water the limiting ones.

With the onset of a lengthy run of below-average rainfall years, beginning with the end of the 1967 rainy season, pastoral populations found themselves occupying very marginal rangelands at the southern edge of the Sahara. The end result was the destruction of the vegetative cover by overgrazing, trampling, and firewood gathering. In the Sahel, soils are extremely fragile and therefore susceptible to wind erosion with the removal of their protective vegetative cover as a result of either overgrazing or for the planting of annual crops. These factors combined to cause desertification throughout the region. With the decimation of their livestock herds, pastoralists fled to villages and cities to the south, where tens of thousands ended up in refugee camps. Thus, farmers and herders had put themselves (with the help of their governments and foreign aid specialists) out on the proverbial limb of a tree. They considered themselves victims of a natural disaster which, as noted earlier, proved not to have been the case.

The end result of these societal and ecological processes has been the creation of desert-like conditions in many parts of the region. This led to the belief that the Sahara was on the march toward the wetter south. In fact, land was being degraded and 'desertified' at the arid margins of the desert. As recently desertified areas became linked to existing deserts and to each other, an impression was generated that deserts were 'on the march,' spreading throughout the Sahelian zone.

Farmers abandoned their bare fields either when crop yields declined or later, when drought conditions persisted from one year to the next, leaving the fields open to wind erosion and dust storms.

> In the past, the supply of land was not a significant constraint on expansion of output over much of black Africa. But increasing population pressure in many areas has led to a lack of fallow land, so that expansion of cultivated areas, in the absence of other means of maintaining soil fertility, carries the risk of declining yields.[13]

Pastoralists lost most, if not all, of their livestock and their livelihood to drought. Those who had managed to save part of their herds migrated southward searching for fodder. This put them in conflict with farmers whose crops were still in the fields in the regions with better rainfall. Many of the less fortunate displaced herders and farmers ended up in urban slums as beggars or in refugee camps being fed by international relief organizations.

During the drought of the early 1970s, a genuine food crisis developed in the West African Sahel. Food production declined sharply throughout the region and the Sahelian governments had to depend on large amounts of humanitarian aid from the international community to supply their citizens with minimal levels of nutrition in order to survive. Compounding the inadequacies of food production systems and processes to meet domestic food needs in the Sahel was the fact that cash crops were generally produced on the better watered, often irrigated, lands.[14] David Campbell suggested that 'increased [food] demand could theoretically be met by increasing the acreage but the best land was already under cultivation for cereals or groundnut and expansion involved the incorporation of less productive land.'[15] It is important to note, however, that Sahelian governments maintained their cash crop production during the drought and that the cash crop exports of several Sahelian countries actually increased during the drought and famine years.[16] A similar situation occurred in war-torn Ethiopia during the 1972–4 drought and famine, when an estimated 200 000 to 400 000 lives were lost, precipitating the collapse of the political regime of Emperor Haile Selassie.[17]

Aside from the climate factor, what had definitely changed in these areas during the century were the human activities. Campbell succinctly underscored this problem: 'as cultivation was extended into less productive land and as population increased, so the region's ability to meet its people's food needs was jeopardized.'[18] As Reynaut once observed:

> the extension of the groundnut area has continued, especially towards the northern margins of the cultivable zone where, due to the introduction of varieties with a shorter growing cycle, it has been able to compete with millet in an area which was formerly the latter's exclusive domain. Millet has thus been pushed progressively northwards into regions which not only have a lower average ... but also a more irregular rainfall and a high risk of drought.[19]

Therefore, the probability that those activities would be adversely affected by the region's 'normal' rainfall variability would increase.

My contention is that researchers may confuse the adverse environmental consequences of demographic changes under 'normal' climate conditions (identified here as drought follows the plow) with the regional impacts in the Sahel of a possible global warming. Until scientists are in a position to produce reliable scenarios about the consequences of a global climate change for the African continent, future catastrophes associated with the recurrence of drought in the West African Sahel can be characterized as a situation in which drought has followed the plow.

In a recent World Bank publication, Ho reported that 'only a very small proportion of potential agriculture land is under cultivation in the region and except in extreme cases ... agriculture is under little pressure to move to marginally productive lands.'[20] However, Ho's definition of 'potential agriculture land' was extremely broad and included 'all land currently under cultivation or planted to tree crops plus permanent meadows and pastures plus forest and woodland.'[21] Such a suggestion to cultivate this 'potential' appears highly unrealistic from an ecological standpoint. The best rain-fed agricultural land is already in production and, wherever people go in the future for the purpose of rain-fed farming, the land will be likely to be less productive. Much of the literature on the West African Sahel supports this premise. Degradation of the land would surely follow attempts at sustained cultivation of what Ho has referred to as 'potential agriculture lands.'[22] During favorable rainfall periods, cultivation of these lands might be productive, but with a return to drought conditions the productivity of the area's farmlands would decline. As Rolando Garcia has suggested, 'droughts *do not* generate the disequilibrium, they merely *reveal* a pre-existing one.'[23]

Somalia

JÖRG JANZEN

The consequences of anarchy and clan warfare in Somalia (Figure 10) in the early 1990s have compounded as well as overshadowed the adverse impacts of meteorological drought that also plague the country's inhabitants. While Somalis have not been strangers to severe, prolonged, devastating drought, they were not at all prepared for the double negative impacts of war and drought. As in other drought-affected African countries facing severe food shortages in the mid-1980s, only in those countries engaged in internal wars did food shortages turn into full-blown famines. Histories of the interactions between these two phenomena in other parts of the African continent have been recorded since the mid-1980s. However, such a history of the interrelationships between conflict, climate, and food security in recent years in Somalia remains to be written. Nevertheless, a look at Somalia's recent (pre-Civil War) past could provide a clearer understanding of how scarce land resources had been used for agricultural production purposes.

About half of the Somali population lives permanently in settled communities, the other half being nomadic pastoralists or agropastoralists (who practice a mix of farming and livestock raising). Agriculture in Somalia can be divided into three subsectors: (1) nomadic pastoralism and agropastoralism, practiced outside the areas of cultivation, especially in northern Somalia; (2) the traditional subsistence agriculture practiced by small farmers (rain-fed and small irrigated farms); and (3) export and market-oriented farming on large irrigated plantations along the lower Jubba and Shabeelle rivers.[1]

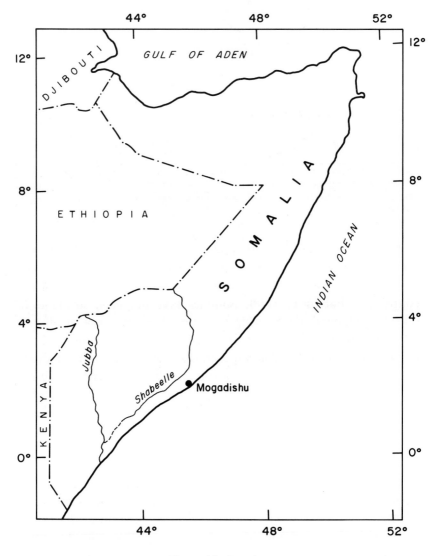

Figure 10. Somalia.

Average annual rainfall statistics (Figure 11)[2] convey an inaccurate picture of Somalia's arid and semiarid climate. The climate is characterized by strong regional, seasonal, and annual rainfall variability. It is mainly this high degree of rainfall variability,

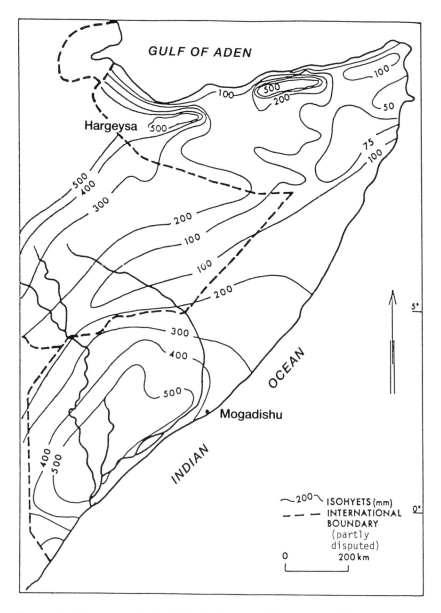

Figure 11. Mean annual rainfall in the Horn of Africa. Adapted from Krokfors.[2]

characteristic of arid lands, that creates serious risks for those involved in rain-fed farming activities, especially those using inappropriate land-use methods.

During the past several decades (and especially since the mid-1970s), the process of desertification has been accelerating throughout Somalia. To a large extent desertification appears to have been the result of the adverse side effects of modern economic development strategies which, in turn, were the consequence of radical changes in the country's political and socioeconomic settings.

The devastating African droughts of the 1970s not only affected the West African Sahel but extended across the width of the continent to the Horn of Africa. The 1974–5 ecological catastrophe, known in Somalia as the '*Dhabaadheer* Drought,' resulted in famine and in hundreds of thousands of nomads losing large parts of their herds and their livelihoods.[3] Considerable attention during that drought focused on the plight of Somali nomads, whereas the heavy economic losses incurred by farmers in Somalia's marginal agricultural lands received relatively little attention from the outside world. There is strong evidence that this drought sparked catastrophic ecological destruction in the marginal agricultural areas, because that land had already been considerably degraded under normal rainfall conditions as a result of inappropriate land-use practices.

The notion of 'drought follows the plow' in a Somali context raises the following questions:
- Which socioeconomic changes in Somalia in recent times can be identified as the most significant with respect to their negative consequences for the country's natural resource base?
- What are the underlying reasons for the extension of inappropriate agricultural land-use practices?
- To what extent do inappropriate rain-fed agricultural practices on marginal lands promote desertification?
- Which regions of Somalia are particularly endangered by desertification?

- Which measures to combat desertification have already been introduced, and which ones would be desirable for the future?

Political and socioeconomic change and its land-use consequences

The existing social structure for major segments of the population in the rural areas, as well as the structure of the Somali economy, has been determined to a large extent by foreign influences. The following factors have contributed to the present conditions in Somalia: Italian colonization in the 1920s and 1930s, the short-lived implementation of socialist development strategies after 1969, the lingering effects of the Ogaden War of 1977-8 and the destitution of hundreds of thousands of political and drought-related refugees, the considerable infusion of development and refugee aid from western international donors after Somalia's break with the Soviet Union in 1978, Somalia's recent heavy economic dependence on the oil-producing Gulf states, and the recent internal war and political changes, followed by a total collapse of a national government replaced by clan warfare.

The rural areas have also been affected by the Somali government's declared policy of settling nomads (i.e., sedentarization). Large projects were actually implemented in 1975 for the resettlement of nomads and, as a result, more than 30 refugee camps were established throughout the country. This policy was an attempt by the government to use the devastating impacts of drought as a reason to settle in agricultural communities and fishing villages large numbers of pastoralists whose lives had been disrupted. Somalia's high population growth rate has also contributed to the degradation of land resources (the population has nearly doubled since independence in 1960, based on recent population estimates of about 6 million).

The development of an export- and market-oriented irrigated agriculture and the commercialization of parts of the rural economy began in the Italian colonial period and has continued to the present. Development efforts have focused on the buildup of medium- and large-scale agricultural operations, especially since the socialist revolution of 1969. The emergence of medium-sized farms, owned to an increasing extent by prominent people such as administrators, military officers, and wealthy traders living in towns, not in the rural areas (Figures 12 and 13), took place in the mid-1980s.[4]

The expansion of export- and market-oriented irrigation farming has had negative consequences for the living conditions of small farmers in many places. Because a shortage of land already exists on the fertile alluvial soils along the lower reaches of the Jubba and Shabeelle rivers, as a result of the expansion of irrigated agriculture, the proportion of irrigable land in the hands of small farmers has diminished, especially near towns and markets. In some areas small farmers have been pushed onto marginal lands in the adjacent savanna zone, where rain-fed farming is at a higher risk to meteorological and agricultural drought. Encroachment on agriculturally marginal areas, which began during Somalia's colo-

Figure 12. Enclosed field in north Somalia. The cutting of trees and bushes for fencing has contributed to the heavy degradation of the vegetation cover.

Figure 13. New farmland on the banks of the Jubba River, south Somalia. Wealthy farmers, rich merchants, and influential government officials, living mostly as absentee landlords in the towns, establish new market-oriented rain-fed, and increasingly, irrigated farms on the banks of the river.

nial period on the lower reaches of the Jubba and Shabeelle rivers, has begun to encompass the middle stretches of both rivers.

Another adverse consequence of this expansion is that large areas of bushland have been cleared for cultivation. Larger tracts of land are needed by farmers to secure a livelihood from rain-fed farming activities than from irrigated agriculture, because of the latter's higher and more reliable inputs (e.g., fertilizers, water) and consequently higher yields. In turn, the living space of the nomadic population, to whom the river valleys and the adjacent savanna zone have served as important grazing lands during the dry seasons, has been shrinking. This shrinking of rangelands leads to a higher density of livestock on the remaining grazing land near the rivers, to a corresponding increase in damage to vegetation and, eventually, to the onset of desertification processes. Such a process of degradation occurs particularly in the old Sandridge zone that runs parallel to the Indian Ocean coast of central and southern Somalia.

Another significant impact of agricultural development in rural Somalia has been the increasing tendency shown by nomads and semi-nomads (agropastoralists) toward voluntarily settling on the land. The transition from nomadic life to a less mobile way of living and, finally, to complete sedentarization accompanied by rain-fed farming has historically been a slow process which has taken place during past centuries in areas with favorable climate and soil in southern Somalia. The sedentarization process, however, has accelerated during the last few decades, for the following reasons: droughts (especially in 1974–5) have accelerated environmental degradation, worsening the living conditions of nomads and agropastoralists. As a result, they have taken up rain-fed farming, become urban traders, or emigrated as workers to oil-producing countries on the Arabian peninsula. Sedentarization has led to changes in the pattern of mobility and distribution of the population, resulting in a concentration of people near cities, roads, deep wells, large water catchments, and other infrastructural facilities in order to farm.[5]

Regional land-use variations and environmental degradation

Across Somalia there are major differences among regional physical conditions because of the country's vast north–south extension. There are also varying levels of importance and dependence placed on farming by different Somali groups. Some examples of inappropriate land-use practices and their adverse impacts on the natural environment in different parts of the country are provided in the following paragraphs.

In *northern Somalia* the dominant activity has been nomadic livestock raising. With the exception of a few areas in northwest Somalia, rain-fed farming has been of marginal importance. However, agropastoralism has become increasingly prevalent, in spite of the fact that the region's climatic conditions make sustained agricultural activities a very risky endeavor. Agropastoralists concentrate their herds close to new settlements and watering points and use small plots of land for rain-fed agriculture, mainly to cultivate either sorghum or grass as fodder for their livestock. Thornbush fences are built to prevent livestock from entering cultivated areas. The cutting down of a large number of bushes and trees for construction accelerates the degradation of vegetative cover, which is already under great pressure from livestock grazing and firewood gathering. Also, furrow systems which have been constructed for gathering rainfall runoff and for transferring it to cisterns or onto farms can initiate as well as exacerbate erosion processes. In Somalia, erosion by water and wind is a very common occurrence in areas where the land has been overgrazed and where rain-fed farming has been developed.

Rain-fed cultivation is concentrated in the *Sandridge belt area* (the Indian Ocean coastal zone). Originally, this region was widely covered with a rich savanna vegetation, but overexploitation by human activities has degraded the topsoil in many places. According to estimates in the 1980s, crop production (mainly sorghum and cowpeas) plays a significant role in this zone, providing about one-third of central Somalia's food requirements. This belt is separated from the sea by a narrow strip of grasslands. Agropastoralism is also prevalent in this area. Significant changes in land-use practices have taken place within the last 20–30 years. As in other parts of Somalia, an increase in population, the commercialization of the livestock sector, and the drilling of deep wells to provide a reliable year-round water supply have contributed to the sedentarization of pastoralists as well as to the expansion of cultivation in the vicinity of new wells.

Local rates of desertification processes tend to accelerate with the construction of deep wells.[6] Within a radius of several kilometers around these wells, nearly all bushes and trees have been cleared and hardly a piece of arable land (or rangeland) remains that has not been enclosed by fences. The large areas prepared for cultivation by bush-clearing and by letting land lie fallow (which is still subject to grazing pressures) are at considerable risk to wind erosion. In some areas fertile topsoil is blown onto the shrublands, while in other areas fertile land becomes covered by sand from nearby sand-dune fields. In yet other areas active sand dunes are rapidly formed.

The development of local pockets of desertification resulting from human activities is not a new phenomenon in this area. The traditional system of cultivating the more favorable humid locations in the valleys between the old sandridges had in the past already resulted in heavy wind erosion, once the natural vegetative cover had been destroyed. Thus, a portion of the shifting sand dunes that present a problem within the cultivated zone today is the direct consequence of earlier inappropriate land-use practices.

The natural increase in the coastal population and, more importantly, the high influx of people into the interior of Somalia in the past two to three decades, as a consequence of prolonged droughts, have increased pressure on the land. In this context large numbers of nomads and agropastoralists from the interior have concentrated their economic activities (of which rain-fed farming is an important one in the coastal zone) on the fringes of the old Sandridge belt and the fertile alluvial floodplains of the rivers for most of the year. This concentration on these margins has led to the degradation and even disappearance of vegetation.

The most important areas of Somalia's rain-fed agriculture are concentrated in the *inter-riverine zone* (between the Jubba and Shabeelle rivers), where sorghum has been the main crop. Here, the population consists of farmers, agropastoralists, and nomads. A major extension of the cultivated area has taken place during the past few decades, because of the local population increase, and the immigration and settlement of nomads who have taken up rain-fed cultivation in addition to keeping livestock. The introduction of mechanization such as bulldozers for bush-clearing and tractors with iron plows has also allowed for, as well as encouraged, a more rapid and widespread extension of marginal land being put into cultivation.

The high demand for farmland in the main areas of rain-fed agriculture (around Diinsoor, Baydhaba, Buur Hakaba, and Wanle Weyn) has led to the merging of isolated bush-cleared patches of farmland into large unbroken cultivated areas. During the dry seasons, these areas are always extremely short of moisture. As a result, strong seasonal winds frequently generate dust storms that carry large amounts of fine, fertile topsoil into adjacent bushland, especially following the harvest.

This set of examples of inappropriate land use in various parts of Somalia shows that desertification processes are common throughout the country. They also point out the interrelatedness of the country's political, demographic, and socioeconomic development and the resulting changes in land-use practices that generate, in turn, adverse changes in the natural environment. Such changes include the intensification of desertification processes and a worsening of drought impacts. In addition, overgrazing by nomadic livestock herds has contributed to these processes. Desertification has also resulted from the extension of rain-fed farming into land too marginal for sustainable agricultural production, such as the old Sandridge zone and adjacent areas along the coastline of the Indian Ocean as well as the inter-riverine area where precipitation is considered more reliable for agricultural production. The combination of rain-fed farming in marginal agricultural areas and the keeping of large numbers of livestock close to settlements has clearly increased the vulnerability of the inhabitants of those settlements to the impacts of a meteorological drought.

The following points are extremely important to keep in mind with regard to future attempts in Somalia to extend agricultural activities into marginal areas in order to intensify crop production:

- Only agricultural projects that are ecologically sound should be undertaken. This suggests that new land must not be opened up to cultivation by clearing the natural vegetative cover from large areas, but by forming strips of cultivated land alternating with belts of natural vegetation. It is vitally important that these strips be diagonal to the main direction of the wind.
- Strip farming should be carried out by small farmers and agro-pastoralists as well as by those who manage the large agricultural projects.
- Large-scale projects based on sophisticated land-use practices borrowed from other countries (Figure 14) do not seem to be

Figure 14. Kurtun Waarey Agricultural Settlement Project, south Somalia. Large-scale, strip-pattern dryland farming systems with expensive sophisticated farm machinery provided by development aid have not yet proved to be an ecologically sound farming system for the semi-arid areas of southern Somalia.

a reasonable alternative for Somalia, given existing climatic, ecological and socioeconomic conditions.
- New farmland should be developed only in such a way that traditional migration patterns by those with livestock are not disrupted. This applies directly to the rapid and uncontrolled extension of irrigated cultivation in the river valleys of the Jubba and the Shabeelle.
- In general, direct government intervention in rural Somalia should be minimized. However, in situations where environmental destruction has already reached a high level, such as in parts of the old Sandridge zone on the Indian Ocean coast, or in the gallery forest zone by the Jubba River, it is necessary to establish an efficient agricultural extension service to prevent the introduction or continued use of inappropriate land-use practices. Rapid, uncontrolled deforestation must be stopped immediately.
- With respect to rural development planning, the relationship of socioeconomic factors to the natural environment (including local and regional climates) should be explicitly considered and taken seriously.

The Brazilian Nordeste (Northeast)
ANTONIO MAGALHÃES AND PENNIE MAGEE

The Brazilian Nordeste encompasses about 18% of Brazil's total land mass and contains about 30% of its population (Figure 15). It remains an underdeveloped part of the country, despite favorable economic growth rates in the last few decades. Per capita income remains low and income distribution is skewed. A large part of the Nordeste population earns wages below the Brazilian poverty line. The combination of large numbers of poor people eking out a living from land marginally productive for rain-fed agriculture and subject to highly erratic rainfall from one year to the next has prompted many inhabitants of the region, known as *nordestinos*, to migrate.

The region is in fact well known for recurrent, often devastating droughts and their major social, economic, and environmental consequences. Severe and sometimes prolonged droughts have been recognized by scientists and policymakers for more than a century as an integral part of the physical and social setting of the region. In fact, the region and its droughts have been the subject of many Brazilian novels. Some of the best known of these include Euclides da Cunha's 'Rebellion in the Backlands' (*Os Sertões*), published in the 1890s, and Graciliano Ramos' 'Barren Lives' (*Vidas Secas*), which appeared in the 1930s.

Regional ecosystems

The Nordeste can be subdivided into five large ecosystems:[1]

Figure 15. Brazil.

(1) The *zona da mata* (coastal zone) is a humid area, with an average precipitation of 2000 mm/year. The main activities are sugar cane and cocoa production on large landholdings.
(2) The *cerrado* area comprises the sparsely populated western part of the Nordeste, where annual precipitation exceeds 1000 mm.

The main economic activity is soybean production, also on large landholdings. Most of the inhabitants are recent settlers who migrated from the southern part of Brazil.

(3) The *pre-Amazonia* is a transition zone between the semiarid area and the superhumid Amazonia. Precipitation is high. In the last 30 years, human and agricultural settlements have encroached on this area. Rice production and cattle raising constitute the main activities.

(4) The *agreste* is a transition zone between the coastal forest and the semiarid zone. Rainfall is about 800 mm/year. Severe drought events in the neighboring semiarid *sertão* usually affect this area. Population density is high. Agriculture (food production) and cattle raising are the main activities.

(5) The semiarid *sertão* is the most underdeveloped part of the Nordeste. With relatively few natural resources, shallow soils, scattered drought-tolerant *caatinga* vegetation and high climatic variability, this area is subject to frequent droughts. Precipitation ranges between 300 and 800 mm/year.

The semiarid area is highly populated: 20 million people live in the *sertão* and the *agreste*. The natural resource and economic bases are insufficient to support the population and, as a result, poverty is widespread throughout the area. Agriculture predominates and consists of food crops (beans and maize) and cotton, together with extensive cattle raising. However, climatic variability brings a high risk to agriculture. Droughts cause crop failures, as well as social problems such as unemployment, malnutrition, hunger, and out-migration.

Drought recurrence

Runs of drought years are not an uncommon feature of the climate of the *sertão*. The neighboring *agreste* is also affected by drought but not as severely, since it is a transition zone between the cattle-raising *sertão* and the better-watered, sugar-cane-producing coastal *zona da mata*. The *cerrado* and the *pre-Amazonia* are much less drought-prone.

Superimposed on this general ecological breakdown is an area referred to as the 'Drought Polygon' (Figure 16). The semiarid

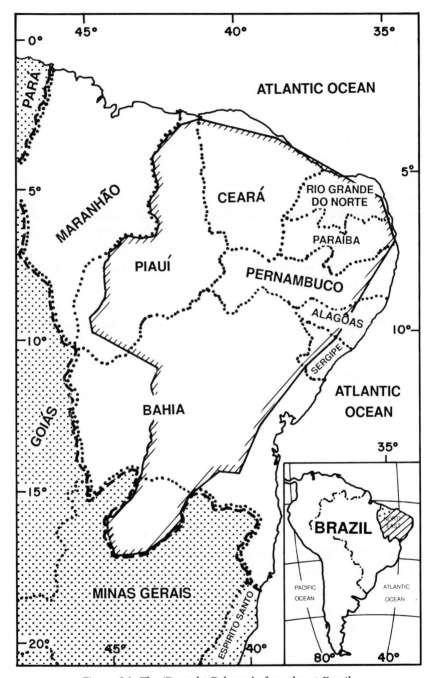

Figure 16. The 'Drought Polygon' of northeast Brazil.

sertão forms the heart of the Drought Polygon, experiencing drought on the order of once every 5 years. The concept of the Drought Polygon was created by the government in 1936 and defines the area in the northeast that was most susceptible to severe climate variations and sporadic drought conditions.[2] This regional concept is political as well as meteorological, as it determines legal eligibility to receive federal funds for drought relief. Because of its political nature, it has been redefined to encompass several political units, making those units eligible as well for drought aid, regardless of need.

Colonial legacy

The present socioeconomic setting and drought impacts in the Nordeste are best understood when placed in the context of a brief description of Brazil's colonial development. Human settlements and economic activities began in the coastal *zona da mata*. Since the sixteenth century, this area has been dedicated to the production of sugar cane. In order to assure the primacy of sugar cane production, economic activities such as subsistence crop production and cattle raising were not permitted.[3] The search for alternative locations for the raising of cattle during the colonial period forced settlers not belonging to the sugar cane elite to move into the semiarid *sertão*.

As a result of the search for land outside of the sugar cane area, the economy that first developed in the *sertão* was based on cattle raising and subsistence agriculture. The region became a source of meat and cattle hides, which complemented the sugar cane economy of the *zona da mata*. Those initially involved in cattle raising in the *sertão* had access to abundant land. Some of the largest estates encompassed thousands of square kilometers. Typically, their land bordered a river and extended away from the river, cutting across less fertile lands. The availability of abundant free land attracted prospective settlers to the otherwise hostile environment of the *sertão*. This area was, however, situated in the heart of the Drought Polygon and, during extended drought episodes, herds would be decimated by the sharp reduction in water and forage.

Cattle that survived the drought became the nucleus for the reconstruction of drought-decimated herds.

Sugar cane production was (and still is) subject to the whims of international market supply and demand as well as to the vagaries of weather at home and abroad in other sugar-producing regions. In the early colonial period, one of the several declines in sugar production prompted the severing of the trade linkage that had developed between the *sertão* and the *zona da mata*. As a result, the economy of the *sertão*, based primarily on cattle raising and subsistence food production, became autonomous. In the early 1860s cotton became a cash crop important to the region's economy. The fact that cotton is somewhat drought resistant also added to the success of its production in this region. Over time, cotton production, cattle raising, and subsistence food production proved to be complementary activities.

Extending agricultural frontiers

Within the *sertão*, human settlements initially began on the best land, situated along river banks and in humid and hilly areas. Today, these areas are densely populated and contain the major share of agricultural production activities. The remaining dry lands in the region are generally encompassed within the large landholdings, considered necessary for extensive (as opposed to intensive) cattle raising because of their lower productivity as grasslands. Hence, they have relatively low population densities.

This brief history of the Nordeste illustrates the interaction between social, economic and political factors, and meteorological drought. The consequences of this interaction continue to characterize the region today. Specifically, agriculture, especially subsistence agriculture, has been pushed into increasingly marginal areas. The following four factors play central roles in this process:

(1) Subsistence crop production often loses out to cash crop production, whenever the two compete for the use of scarce land and water resources. For example, the most productive lands in the coastal *zona da mata* have been devoted to the production

of sugar cane on large landholdings. In the 1970s, regional cane production received a major boost as a result of fuel shortages that accompanied the first Middle East oil crisis. Alcohol produced from sugar cane was seen as a potential domestically produced substitute for high-priced imported petrol. The loss of this agricultural land to cane production was at the expense of food production.

(2) There has been an increase in competition for agricultural land from those who prefer to raise livestock, primarily because it is an activity that has received government fiscal incentives from SUDENE (the Superintendency for the Development of the Nordeste), during the past few decades. Using incentives such as subsidizing the cost of production, some of the best agricultural lands have been converted to pasture.

(3) The population increase in the Nordeste has driven up the need for food, not only for inhabitants of the rural areas, but especially for those in regional urban centers as well.

(4) Declining soil fertility on 'tired' overworked lands has had a negative impact on crop yields. Losses resulting from declining yields must then be made up by expanding the land under cultivation into as yet uncultivated areas, e.g., the agriculturally marginal lands.

Land ownership in the Nordeste is a key factor in the ability of different groups in society to withstand the impacts of multiyear drought. Ownership of land takes two forms: (1) large landholdings and (2) small landholdings (Table 1). The large landholdings encompass both productive and marginally productive lands, both of which are characteristically underutilized. The soils of the small landholdings tend to be overexploited and often exhibit signs of decreasing fertility.

The scarcity (or, more accurately, the nonavailability) of arable land is primarily a socioeconomic problem. Potentially productive land does exist within the boundaries of many of the large estates, but the owners of those estates have been reluctant to put that land into the production of food crops. They reason that, should drought recur, the land would be required for grazing their cattle.

Table 1. General distribution of landholdings in Northeast Brazil as of 1980 (after Magalhães and Rebouças)[4]

Size of farms (ha)	Percentage of total number	Percentage of total area
<5	56	3
5–20	22	6
20–100	16	19
100–500	5	28
500–1000+	1	44
Total	100	100

Thus, subsistence level food production takes place mainly on small landholdings or on small tracts of land leased to poor farmers within the large estates.

Coping with drought

The ever-increasing human and livestock populations in the *sertão* have made the subsistence economy of the region vulnerable to single-year droughts, and especially to successive drought years. As noted a decade ago,

> The people, the crop varieties and the farming practices have evolved to cope with the harsh conditions ... With little more than a hoe and leftover seeds from the previous year, [the *sertão* farmer] produces subsistence crops (corn, beans, and manioc) for his family and a cotton crop to purchase minimum necessities. Even during years of good weather, the small-farmer populations in this region are at risk because of low agricultural productivity. Drought episodes, especially multiyear ones, upset the precarious balance between subsistence food production and food intake.[5]

During the most recent multiyear drought in the Brazilian Nordeste between 1979 and 1983, newspaper headlines were once again filled with stories about its impact on people, livestock, and land. With so much attention having been focused on drought in the region since the mid-1800s, one would have expected that by the end of the twentieth century the numerous drought mitigation actions taken by various levels of governments, past and present, would have resolved many of the drought-related problems in the Nordeste. This has not occurred, however, in part because of two important trends in the past century: (1) the increasing dependence on marginal lands for the production of subsistence crops and (2) the types of drought mitigation policies that governments have pursued.

With respect to an increasing dependence on the use of marginal areas within a region that is already considered marginal to agricultural production, the expansion of agricultural activities and livestock production has made the soils and existing vegetation extremely vulnerable to climatic variability. This can be witnessed by the persistence of extremely low, even declining, levels of productivity (i.e., yields). When development does occur on the large estates, it first takes place on the relatively more productive land and later on the poorer, more vulnerable areas. In addition, population increases in the Nordeste intensify the need for an increase in the production of subsistence food crops not only for the highly vulnerable rural populations but for food and raw materials for the urban dwellers as well. These pressures, in turn, lead to the need to put more of the marginal lands into cultivation and to reduce fallow time on the good lands, which thereby reduces the quality of the soils, and hence leaves them increasingly vulnerable to meteorological and agricultural drought (Figure 17). As an Earthscan report once noted, 'the concentration of land in the hands of the rich in the huge Northeast section has forced peasants to overcultivate their small patches.'[6]

With respect to government intervention, there has been a reliance in the past on technological fixes such as the construction of large and small dams, wells, transportation, and communications infrastructure. There has also been an increasing emphasis on irrigation projects. Lands put into irrigation in the region, however, encompass a relatively small percentage of the total agricultural area. Furthermore, irrigation requires a large initial investment

Figure 17. Degraded marginal lands in the State of Ceará (Northeast Brazil).

and a continual supply of costly inputs such as fertilizers, improved seed varieties, and water. Resorting to irrigation, therefore, is not a viable option for subsistence farmers.[7] While more productive with respect to increased yields, irrigation has not eased subsistence food production problems, since irrigation is most frequently used to produce cash crops such as sugar cane and cotton. Such crops are favored because they can bear the cost of the required expensive inputs. Such projects, geared toward the irrigation of cash crops, have led to the expulsion of farmers of small plots, causing them either to encroach on increasingly less productive areas or to seek employment in distant urban areas throughout the country.

Regardless of their ideological stance, the various Brazilian governments have resorted to societal as well as technological fixes. These include work programs for those who had become unemployed as a direct or indirect consequence of the impacts of drought (Figure 18). Farmers temporarily out of work have been paid as

Figure 18. Drought-plagued farmers temporarily employed on public works projects in Paraiba.

laborers on public works projects. This is a subsidy from the federal government to the region's unemployed in part to keep *nordestinos* from migrating to major cities, such as São Paulo, in search of wages. This has been one of the popular tactical political responses by recent governments to the chronic droughts (or, more importantly, to the drought-induced out-migrations) that plague the Nordeste. However, drought-related work programs have failed to address basic underlying social and economic problems associated with land tenure and agricultural production in the region.

Official drought responses

In a recent assessment of climate impacts related to government strategies in response to climatic variations in the Nordeste, the authors categorized the development of drought policies in the following way: (1) the study phase, 1877–1906; (2) the engineering (and water supply) phase, 1906–45; (3) the ecological phase, 1945–50; (4) the economic development phase, 1950–70; (5) the socioeconomic development phase, 1970–90; and (6) the sustainable development phase, 1990 to the present.[8]

(1) *The study phase (1877–1906)*. This phase is characterized by a series of reports and a few extemporaneous actions. After the catastrophic 1877 drought, a National Commission of Inquiry was established by the Emperor of Brazil to study the drought problem and recommended the construction of water reservoirs, the drilling of wells, and the construction of canals, roads, and ports. During extreme droughts, the government provided limited drought relief, consisting mainly of food distribution, with almost no positive results. Incentives were created for people to move to the Amazon to work as rubber tappers, as an escape for the increasingly desperate population of the Nordeste.

(2) *The engineering and water supply phase (1906–45)*. In 1909, a permanent institution – the Superintendency for Studies and Drought Relief Works – was created. Since 1945 it has now been known as the National Department for Drought Relief Works (DNOCS, in Portuguese), charged with addressing the problems generated by Nordeste droughts. During this phase, DNOCS emphasized building a water supply infrastructure that today stores more than 22 billion cubic meters of water. However, there was no corresponding provision for the large-scale use of the stored water for farm production activities and irrigation.

(3) *The ecological phase (1945–50)*. During this brief period there was an effort to implement a strategy to strengthen farm production by introducing drought-resistant crops. Agriculture was to be adapted to the ecological conditions of the semiarid region. Farmers were advised to plant hardy varieties of cotton and other crops.

(4) *The economic development phase (1950–70)*. Following the trends of industrial development in Brazil, there was a widespread conviction in the 1950s that, in order to reduce regional disruptions caused by recurrent droughts, a strong

agricultural sector had to be complemented by industrial development. SUDENE became the region's paramount development agency. However, after 1964 a new federal policy reduced SUDENE to a shadow of its former self. In 1970, the following new priorities were set: (a) the integration of the Nordeste into the national economy, including into the Amazon region, and (b) public irrigation. The Transamazonian Highway was built in part to stimulate migration from the Nordeste to the Amazon, and irrigation projects were implemented in the Nordeste to foster production and employment for those who decided not to emigrate.

(5) *The socioeconomic development phase (1970–90)*. During the first half of the 1970s, it was realized that to cope with droughts, attention had to be paid to the region's glaring social problems. They had to be addressed over and above meteorological, 'ecological,' economic, and development factors. In marked contrast to previous strategies to create or adapt institutions, the policy response to this enhanced dimension of Nordeste drought was to establish integrated rural development strategies through projects to be executed by existing institutions. Several projects have been designed and executed in the last two decades.

(6) *The sustainable development phase (1990 to the present)*. Environmental concern now involves a broader set of considerations. It is now recognized that past development actions have been undertaken to the detriment of natural resources and of future generations. Sustainable socioeconomic and ecological development is now advocated as the proper answer to reduce the vulnerability of the Nordeste population to future droughts. The adoption of sustainable development strategies, although accepted as a political and social necessity, poses a challenge to policymakers and to society, since it is greatly dependent on nonexistent or nonavailable technologies and financing.

Each phase (i.e., approach) has some proponents today. While there is no clear line of demarcation separating these policy phases and while there is overlap among them, one can identify a predominance of interest in the latest approaches that address the social causes and consequences of drought in the Nordeste holistically. Each new development strategy generates a renewed hope among *nordestinos* that it will finally provide them with the optimal economic solution to sustainable regional development.

Recent government strategies have managed to avert the major human disasters like those that used to occur during droughts in the past. The drought-affected population today has access to government relief programs during extended, severe droughts; these programs have without a doubt saved thousands of lives. Changes in water resources and transportation infrastructure have also played a major role in mitigating the impacts of droughts. In past droughts the affected population had little access to water resources and no way to emigrate except to go hundreds of miles on foot. These improvements, however, cannot hide the fact that overpopulation and economic overexploitation, and a resultant incorporation of marginal lands into agricultural production, has made poor people in the semiarid region more vulnerable to the impacts of future droughts. Paradoxically, while the ability of federal and state governments to respond to drought has improved, the vulnerability of ecosystems and inhabitants in the Nordeste has increased. The end result has been that, nowadays, even rather minor meteorological droughts can have major societal and ecological impacts, whereas in the past only major droughts became major problems. In other words, it seems that previously there was a greater resilience to small climate variations.

It is becoming increasingly difficult to separate the impacts of a meteorological drought from those of socioeconomic factors. A decade ago, for example, many *nordestinos* believed that the Nordeste had been plagued by a 5-year drought from 1979 through 1983. The societal impacts of that perceived multiyear drought were major and were measured by agricultural and socioeconomic indicators. Brazilian meteorologists, however, noted that according to meteorological data (e.g., rainfall), there were only three years of meteorological drought during that 5-year period. Nevertheless, political, economic, and social decisions were being made as though a 5-year extended meteorological drought had occurred.

The Nordeste continues to be plagued by the problems generated by an increasing population trying to sustain itself on a dwindling resource base (i.e., increasingly marginal areas). Societal and technological fixes have in the past provided only 'band-aids' for what are clearly more basic underlying social (and political) problems. Such fixes, while seemingly effective in the short term, may serve only to increase the total sum of human misery in the region in the long term.[9] Brazilian governments need to strive for sustainable development in the Nordeste.

Postscript: the Nordeste–Amazon connection

The process of moving agriculture onto marginally productive lands in the Nordeste has its counterpart in the Amazon. Historically, *nordestinos* have moved out of the region altogether to the other parts of Brazil including the Amazon, in response to climatic and socioeconomic conditions at home. However, a significant body of literature written over the past two decades has demonstrated that land in the tropical rainforests of Amazonia is quite fragile and will turn into marginal agricultural land in a very short time if improper agricultural techniques are used in inappropriate landscapes. Hence, *nordestinos* who have migrated to the Amazon typically find themselves once again trying to eke out a living on unproductive land and in the process are unwitting actors in aggravating the problems of land degradation and, through deforestation, regional if not global climate change.

The northeastern and northern (Amazonian) regions of Brazil have a history of interaction that dates back to colonial times. Because of ocean currents and prevailing winds, it was easier to communicate with Portugal from these regions than to communicate with the southern regions of Brazil. Hence, these two northern regions became a political and economic unit distinct from the rest of the country.

For the purposes of our argument here, however, we will focus on the connection between the Nordeste and Amazonia since the turn of the century. There have been four key moments in which people from the Nordeste migrated to the Amazon in response to both 'push' and 'pull' factors. The direction of the flow has tended to be from the Nordeste to the Amazon; the Nordeste has had too many people for too little available productive land, while the Amazon appeared to have too much land and not many people.

The first twentieth-century migration occurred during the peak of the rubber boom in the Amazon around 1900–10. Since the Amazon was sparsely populated, there was a shortage of local labor to tap the rubber trees. During the same period, the Nordeste experienced one of its severe droughts. People from the region moved into the Amazon to become rubber tappers, as an escape from the drought back home. These *nordestinos*, especially from the Brazilian states of Ceará, Maranhão, and Piaui, were assimilated into Amazonian culture from the states of Pará to Acre and became an important element in the formation of the *'caboclo'* culture of the Amazon. Today, in the western states of Amazonia, especially Acre, the term *caboclo* is nearly synonymous with the rubber tappers descended from *nordestinos*.

A second wave of migration from the Nordeste to Amazonia occurred during World War II. The rubber boom had come and gone, leaving Amazonia in a state of economic decay. However, when the sources of plantation rubber in Malaysia were cut off from the Allied forces, the United States turned its attention back to native rubber production in Brazil. The United States and Brazil signed a series of agreements called the Washington Accords, which included provisions for significantly expanded rubber production in Amazonia for the war effort. The two nations embarked upon a strategy that included active recruitment of *nordestinos* to resolve two problems at once: the shortage of labor in Amazonia and the chronically drought-affected, poverty-stricken populations of the Nordeste.

The third wave took place during the 1970s. Once again, the Nordeste was suffering from drought and the Amazon was sparsely populated. The context for this migration differed somewhat from the two previous ones, because Brazil was under military rule. While questions of national security had always been a concern to Brazilian governments, under the military it was of prime importance. Much has been written about the policies implemented in the name of national security during this period, but for our purposes one stands out: the plan to build the Transamazonian Highway. There are several reasons why this highway was built, but to the government it was part of a strategy to occupy the 'vulnerable' Brazilian Amazon region – vulnerable because the Amazon basin is shared by several countries, and because it was only sparsely populated by Brazilians.

The Transamazonian Highway was originally conceived to include planned colonization projects along the length of the road. Between 1970 and 1973, the government and private enterprises actively recruited *nordestinos* as well as the landless poor from Brazil's southern states to live in these colonies, promising them 100-hectare plots of farmland. By 1974, the government shifted its policies to encourage large landowners from the south to invest in the Amazon; the dynamics for the small-scale *nordestino* farmer changed dramatically. *Nordestinos* were left with virtually no support to eke out a living in difficult conditions. Most were forced into living conditions reminiscent of their lives in the Nordeste: to cultivate land for a couple of years and move onto 'new land' when the poor soils were further depleted of nutrients or when these poor farmers themselves were expelled, usually through violent means, by landowners seeking to increase the size of their landholdings.

Currently, there is a fourth wave of migration of small-scale farmers to Amazonia. The 'push' factors which expel these agriculturalists from their native northeastern region remain the same: concentration of land ownership in the hands of a few, and the recurrence of meteorological drought. The 'pull' factors now include the large-scale development projects implemented by the federal government in the 1980s, such as mining concerns. *Nordestinos* continue to harbor the ever-present hope that they might be able to obtain a legal title to a small piece of arable land. The Amazonian state of Pará is especially attractive to *nordestinos* because it is a player, along with the northeastern state of Maranhão, in an ambitious aluminum mining project that ships the raw materials from Pará by train to the capital city and port of São Luis, Maranhão. Since the trainline also carries passengers, it has become a fundamental conduit through which *nordestinos* come to Pará in search of a new life. Thus, it can be shown that no part of Brazil escapes the impacts of recurrent drought in the Brazilian Nordeste. The consequences of drought following the plow in one region clearly affect land use, human activities, and environmental quality in other regions as well.

The dry regions of Kenya

DAVID CAMPBELL[1]

In a book on the Earth's problem climates, the spatial distribution of rainfall deficiency in tropical East Africa, encompassing Uganda, Kenya, and Tanzania (Figure 19), was identified as 'the most impressive climatic anomaly in all of Africa ... Nowhere else in a similar latitudinal and geographical location does there exist a widespread water deficit.'[2]

The natural risk of drought and environmental degradation in Kenya generally depends on the amount, intensity, timing, and distribution of precipitation during the growing season. Many parts of Kenya have two rainy seasons: one around March–May, and the other in October–December (Figure 20). Historical as well as climatological records indicate that drought has been a recurrent phenomenon in various parts of Kenya,[3] with the most recent severe droughts having occurred in 1984 and 1992. Table 2 provides an overview of droughts and their impacts in Kenya since World War II.

In many respects, Kenya's development problems are similar to those faced by other sub-Saharan countries: high population growth rates, highly variable interannual rainfall, limited agricultural land, environmental degradation, ethnic rivalries, rural–urban migration, and unemployment.

Kenya is widely recognized as a prime example of a country where rapid population growth is resulting in land pressure and environmental degradation. As in other countries where population

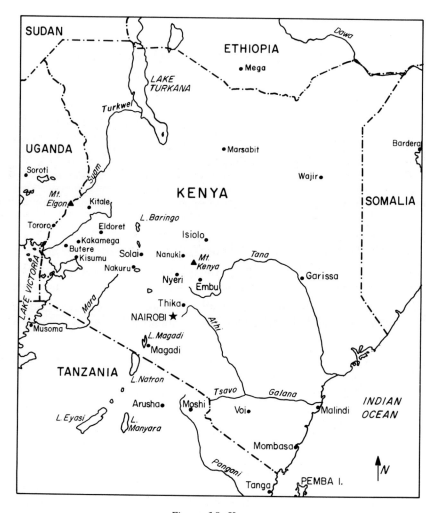

Figure 19. Kenya.

growth is rapid and resources are degraded, the relationship between these processes is neither simple nor deterministic. Rather, the link is complex and is rooted in long-term patterns of interaction between ecological, political, socio-cultural and economic circumstances.

During the 1970s the population growth rate in Kenya reached over 4% a year (Figure 21).[4] In 1990 Kenya's population was

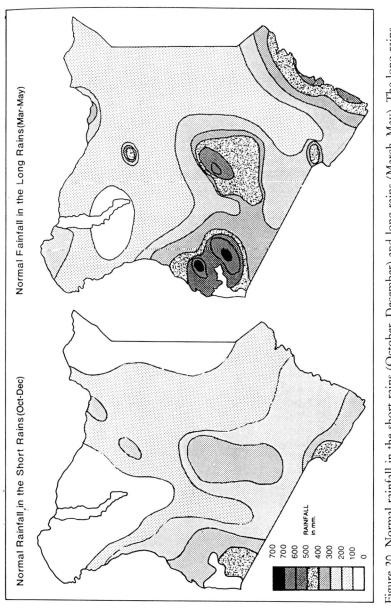

Figure 20. Normal rainfall in the short rains (October–December) and long rains (March–May). The long rains are generally wetter and more reliable, except in parts of Eastern Province. From *Coping with Drought in Kenya: National and Local Strategies*, ed. T.E. Downing, K.W. Gitu & C.M. Kamau. © 1989 Lynne Reinner Publishers, Inc. Used with permission of the publisher.

Table 2. Droughts in Kenya
(modified from Downing et al.)[6]

Years	Region	Causes and effects
1947–50	Kikuyu lands, coast	Rains failed; good harvest in Nyanza and Rift Valley provinces. One of most severe famines at the coast
1952–5	Kitui and other districts	Drought reduced exports by 20–25%
1960–1	Widespread: Maasai lands, Machakos, Kitui, Rift Valley, northern districts	Drought followed by floods caused famine. US aid sent. Cattle mortality among Maasai 70–80%. Widespread crop failure. Ten million Kenya pounds spent on food relief in 1961
1974–6	Kajiado, Kitui, lower Meru, Machakos, Tana River, Turkana	Government projects (food storage, drought-resistant crops, livestock improvement schemes) averted worse effects. Maasai cattle losses as high as 80%
1981	Eastern province	Known as 'famine with cash in my pocket.' Large imports required, partly due to depletion of maize reserves by early exports
1983	Coastal hinterland, Kitui, Machakos, Meru Kakamega, Nyanza	Poor long and short rains; high prices in local markets, some water shortages; cattle and human migration
1984	Central, Rift Valley, eastern and northeastern provinces	Severe drought in long rains over all Kenya except western and coast provinces. Large food imports averted famine. Short rains normal
1992	Nationwide except parts of coastal and western regions	Large-scale food shortages. Minimum famine due to large food imports

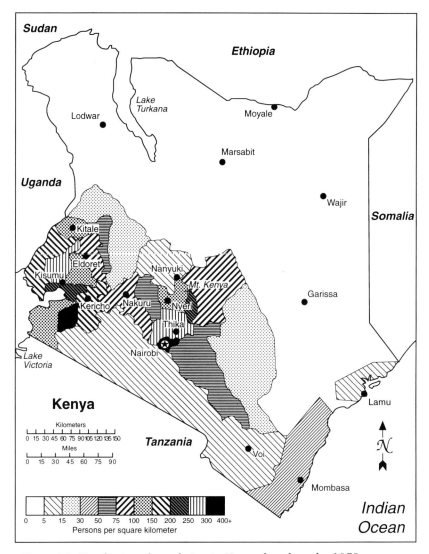

Figure 21. Distribution of population in Kenya, based on the 1979 census.

estimated at 24.2 million, growing at 3.8% a year.[5] It is still among the highest in the world. At such a high growth rate, the country's population can be expected to double in less than 20 years. More than 75% of the people are involved in agricultural production and

75% of farms cover less than 2 hectares. As the population has grown, the land has been subdivided among family members to the extent that many farms today are unable to support the people dependent upon them.

Issues of land use and land distribution are of salient political concern in contemporary Kenya. As Norman Miller noted: 'land and inequality are intertwined and they are the driving forces behind Kenya's current politics.'[7] Their origins lie in the colonial era when European settlers appropriated large areas of productive land in the Central Highlands and the Rift Valley. The impact on the indigenous farmers was immediate and long-lasting. The colonial settlers not only appropriated land from the Kenyan farmers and herders, but they changed the economic structure such that land replaced cattle as the accepted measure of wealth, security, and status.[8] With the increased value of land, demand for it increased at a time when the availability of land of high agricultural potential had been sharply reduced by colonial settlement. Also the African land-use system was based upon a strategy which included regular fallowing to replenish soil fertility. Letting land lie fallow for varying periods of time enables the soil to become replenished with nutrients and its structure to be maintained. Farmers tend to reduce fallow periods when under pressure to produce more food or cash crops. Deprived of measures to preserve or enhance soil fertility (inputs such as fertilizers or irrigation are too expensive for the small farmer), the productivity of the land declines to the point where the farmers cannot produce enough food to support their families.[9] In the absence of investments to improve the productivity of agricultural lands, the continuous cultivation is indeed likely to stress the land base.

The process continued after independence. Most of the settlers' lands were taken by Kenyan elites[10] and today 'an overwhelming number of Kenya's political elite have invested in agricultural land,'[11] though some estates were subdivided to settle the landless. The pattern of shortage of productive land for the majority remained, however, and increased as the population grew. For much of the 1960s and 1970s Kenya was among the most economically successful countries of Africa. Economic success did not, however, result in sufficient investment either in peasant agriculture, in labor-intensive industry to expand the opportunities for rural production, or in urban industry to absorb the increased demands for employment from the rapidly growing population.

Economic gains were unevenly distributed among the people and between different areas of the country. Wealth became concentrated in the hands of a small but politically powerful elite who viewed ownership of land as conferring status and independence. This group used its position to acquire control over large areas of Kenya's agriculturally most productive land. For the majority of Kenyans, however, the ability to acquire land and gain a living from it is limited by the activities of the wealthy, who have obtained much of the best land and who have used their political power in the distribution of government resources.

Both during the colonial period and since independence, the large-scale farms have emphasized production for the market. Coffee and tea, and, to a lesser extent, flowers and vegetables, are grown for export, and wheat is produced for the domestic market. The social and ecological distribution of cash crop production has changed over time. Michael Lofchie has noted that the shift in political leadership from Kenyatta to Moi resulted in a change in emphasis in cash crop production to favor the crops grown in the areas where the elite associated with Moi owned their land. He wrote that 'whereas the Kenyatta government had its economic roots among the country's export-oriented coffee farmers, the Moi administration finds its support among the country's grain growers. Its particular interest has been to stimulate the growth of a Kenyan wheat industry in the hitherto less developed portions of the Rift Valley.'[12]

This shift in emphasis from coffee and tea production in the Highlands in the Kenyatta era to wheat production in the Rift Valley under Moi reflects the changing ecological basis of land ownership over time among different socioeconomic groups. The Kenyatta elite came to power at independence and obtained the former European lands in the areas of highest agricultural potential. By the time Moi's associates came to political power most of this land was already owned so they began to acquire the 'next best' lands. These have included the areas where the World Bank sponsored wheat production in the Rift Valley and also previously forested land where additional tea estates have been created.

The expanding control of the lands of highest agricultural potential by the elites continued the process of disruption of smallholder agriculture that was initiated by colonial settlement. As the smallholder population increased, it has had to seek lands even farther down the ecological gradient, beyond the zones suitable for rain-fed agriculture. As Bernard and colleagues have recently noted: 'population on this small island of good land is building rapidly. Rural migrants are being forced out of the highlands into marginal arid and semiarid regions to the east and south.'[13] The agricultural frontier expanded into previously uncultivated, usually drier, areas. Until recently, these dryland areas have been viewed by policymakers as 'marginal lands.' They received little investment of government funds beyond a few irrigation schemes and funds needed for the management of the national parks which had been set aside for the conservation of wildlife and which generate a significant amount of foreign exchange revenue from tourism.

Over the past 30 years, the agricultural 'frontier' has reached into the rangelands, which are the traditional homes of nomadic livestock herders such as the Boran, Maasai, Samburu, and Turkana. These areas have historically had relatively sparse populations due to the low and variable rainfall. As land of higher potential became closed, the drylands provided the only alternatives for people from farming communities to acquire land. A number of researchers have examined this process in detail. For example, Campbell wrote of Kajiado District:

> the farmers moving into the drier areas have been quick to select the most promising locations – the river valleys and more humid hillslopes – but they remain vulnerable to environmental factors and, as they are recent arrivals in these harsher environments, they have yet to implement cropping and conservation practices which reduce the risk of crop

failure. Similarly, they have yet to develop adequate social linkages to replace those which they broke in moving from their home areas. Thus at least in the initial years of farming in semiarid areas the majority of farmers are extremely vulnerable to the effects of drought.[14]

Peter Little and his colleagues, writing of Baringo District, described the process as follows:

> Landowners in the hills began to grow high-value export crops, such as coffee and pyrethrum, pushing up the price of land and forcing poorer farmers into more marginal lands that were not under private ownership. In these areas, which were often used seasonally by herders, migrants tried to grow food crops, such as maize and finger millet.[15]

The encroaching farmers have also had to confront the existing land users – herders – who had already seen their access to resources reduced by the demarcation of large areas as national parks and reserves for the conservation of wildlife. Since colonial times the governments of Kenya have tried to alter the land-use systems of herding societies. Herders were often characterized by the British administration as uncivilized. For example, the author of the Maasai District Annual Report for 1921 wrote that: 'There can be no doubt, I fear, that the Masai [sic] are a decadent race, and have only survived by being brought under the protection of British rule... They remain primitive savages... They live under conditions of incredible filth in an atmosphere of moral, physical and mental degeneration.'[16] Further, they were accused of degrading their rangelands as a consequence of an irrational cultural affinity with their animals, referred to as the 'cattle complex,' which prevented them from managing their production system in a manner which would sustain the rangeland resources. Close examination of the historical record suggests that the colonial administration recognized that this was not true, even while it used the notions of overstocking and overgrazing to justify its policies of developing the rangelands.

Misunderstanding of the water and grazing needs of pastoral communities was also evident in the demarcation of areas as national parks and reserves. A successful wildlife reserve must enclose sufficient grazing land and water to enable the wildlife to survive dry seasons and periods of drought. Migrations into adjacent areas occupied by herders during the rainy season were not seen as a potential problem when the parks were demarcated. Herders, therefore, had to bear the costs of losing access to specific dry-season water and grazing resources and had to share the resources in the wet-season wildlife dispersal areas.

Competition between and among farmers and herders over the limited areas with secure water supplies and good soils is common, and both groups are questioning the preservation of vast tracts of land for the conservation of wildlife.[17] As farming has increased in areas adjacent to parks, crops too have become vulnerable to destruction by wildlife. The issues of farmer–herder conflict frequently require formal resolution by district authorities. Wildlife competition is more complicated and politically sensitive, because of the large amount of foreign exchange it generates. Since the 1970s, wildlife managers have attempted to develop and implement a strategy which would encourage people living in areas adjacent to wildlife parks to be more accepting of wildlife. The most important element of this strategy has been to return some of the revenues from wildlife viewing activities (such as safaris) to adjacent landowners as compensation for the losses they incurred because of wildlife damage.[18] Initially, these wildlife utilization fees were given to county councils, but this failed to provide the political benefits of greater popular support for wildlife conservation policies because the affected landowners were not directly reimbursed for their losses. Recent changes, however, have led to more direct payments, and where these are made the fees represent a major addition to household income.

Thus, despite being seen by policymakers as having little economic value, these marginal areas have had to absorb increasing numbers of people. They are the contemporary locus of a process of competition over land which has its origins in the colonial period. Smallholders have been forced down the ecological gradient to seek a livelihood on drier lands which offer only an unpredictable potential for agricultural production. There, they deprive herders of access to the better-watered margins of the rangelands and undermine the viability of herding systems already disrupted by government policies on national parks and reserves and regulations concerning livestock movement and livestock prices. A result is that the poor in Kenya's drylands are increasingly being marginalized, both economically and ecologically, and poor farmers compete with poor herders for remaining resources.

The contemporary shortage of land with agricultural potential has not stopped the wealthy from seeking new lands. The economic and social value of land continues to drive the wealthy to seek new locations in which to invest. The drylands have now become the focus of land investments. To yield a return they cannot remain 'marginal' and thus processes have been set in motion to encourage those activities which do have potential economic value, namely cattle ranching and returns from tourism based on wildlife viewing.

While the drylands of Kenya, in general, do not have significant economic activities, there are some areas which have considerable value and potential value. As discussed above, in some locations adjacent to national parks, wildlife utilization fees are now providing some landowners with considerable revenue, and others have invested in tourist lodges and 'cultural bomas,' villages in which guided tours are offered and cultural artifacts are sold. Another source of wealth is livestock raising for meat production, rather than subsistence. The growth of cities in Kenya has been rapid in recent years, and 24% of the population is now living in urban areas. The consequent demand for meat has increased the prospects of economic returns to investment in beef ranching in the rangelands. It is economic opportunities such as these that local people see as attracting the wealthy elites to invest in land in these areas.

The ability of the elites to purchase lands in areas formerly under very strong traditional patterns of ownership is being facilitated by contemporary changes in these patterns. Land ownership in the rangelands was dominated by legal title held by groups that have traditional claims to the land. During the 1970s, large tracts of Kenya's rangelands were formally demarcated and legal title given on this basis. Increasing encroachment of competing land uses and the growth of the population has led to a situation where the economic viability of the group-owned areas is in question. Existing land-owning groups are deciding to subdivide to small plots held by individuals. This process has been encouraged by the Kenyan government. It is extensive, for example, in Kajiado District. Many of the resulting plots are too small to support livestock-raising and are too dry to support rain-fed agriculture. The economic options for their owners are restricted, and many are choosing to sell their lands.

Surprisingly, the demand for these formerly 'marginal lands' is high. As Campbell and Olson have noted:

> Astute politicians and economic elites will seize the opportunity to purchase the individual ranches (called IRs) as a means of gaining access to the potential revenues from wildlife-related activity and beef ranching. It has been estimated that by the year 2010 Kenya will become a net importer of beef and thus the potential economic returns to beef production in areas now under subsistence dairy production are great.[19]

The result is a continuation of the process of 'economic, political and ecological marginalization of Kenya's poor, already realized in areas of higher agricultural potential, . . . being extended down the ecological gradient into the semi-arid rangelands.'[20] Thus, while rapid population growth sets the stage for potential land shortages in a country in which agriculture is the dominant economic sector, the intensity of the process of degradation or sustainability is determined by political and economic conditions.

A brief description of ecological changes in Kenya's Machakos district since the turn of the century is instructive: 'Large parts of the district were virtually uninhabited at the beginning of this century. Yet, by the 1930s, substantial areas were so degraded by agricultural activities that it was thought to be on the edge of ecological collapse.'[21] Summarizing recent studies of this district, David Norse cited several factors that he considered to have averted the collapse and mitigated human impacts on the land: internal migration to more marginal lands, reduced fallow, improved land-use practices, the expansion of cash crop production, and the need for wage-earning workers, among others.

The Kenyan case demonstrates that analyses which merely relate population growth rates to ecological deterioration provide only partial and misleading explanations which can yield only flawed remedial policy initiatives. The evidence from Kenya, and elsewhere, demands a broader analytical framework which recognizes that environmental issues are socio-political issues and that socio-political issues can have environmental consequences. Sustainable and degraded environments arise from complex interactions over time between political, economic, social and environmental processes – interactions the outcome of which is determined by the exercise of power.[22]

In summary, increasing population numbers and the extension of cultivation, pastoralist activities, and human settlements into the recognized drought-prone areas have introduced new problems of land degradation and a heightened susceptibility to the adverse societal effects of meteorological and agricultural droughts. Whatever the causes of extending human activities into marginal areas may be, the notion that 'drought follows the plow' can be used to describe the results of such extensions in Kenya.

Power exercised in the interests of the powerful in countries such as Kenya is resulting in the political, economic and ecological marginalization of the majority. Strategies for sustainable development will have to recognize this fundamental fact and consider its implications for effective implementation of alternative approaches to development.

Australia

R. LES HEATHCOTE

Although the most recent Australian drought took place in Queensland and New South Wales (Figure 22) in 1992–3, the last *major* one occurred in 1982–3. It reminded government and farmers alike of the continuing vulnerability of the agricultural sector of the national economy to fluctuations in seasonal rainfall. Estimates of the economic costs of the 1982–3 drought are shown in Table 3. The social and environmental costs of drought, however, have always been more difficult to evaluate, certainly in monetary terms.[1]

Research into environmental hazards such as drought has shown that adverse impacts on the community occur when there is an imbalance at the interface between human activity (as the 'demand') and natural event systems (as the 'supply').[3] Inadequate or excessive supply (as energy or moisture availability, for example) may cause adverse community impacts. In the case of the 1982–3 drought in Australia, the imbalance was between the moisture needs of domestic crops and livestock feeds and the inadequate supply of seasonal rainfall.

Given the long history of drought in Australia[4] and the implications of research on natural hazards, the question arises as to why the agricultural sector in Australia continues to be so vulnerable. Why have we not been able to buffer our activities against drought over the years? Technological innovations have reduced some of the impacts of drought, but they have not reduced the hazard, and there is some evidence to support the contention that vulnerability to drought may have increased over the years.

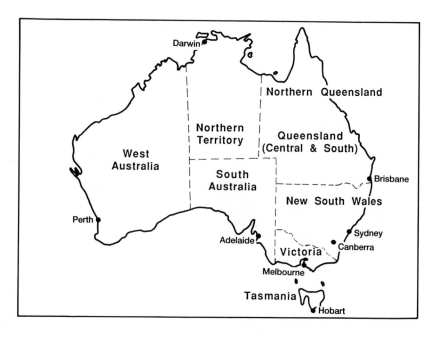

Figure 22. Australia.

First, not all technological solutions have been accepted by the agricultural communities. While some farmers in the semiarid areas experiment with minimum or zero tillage, others still burn their grain stubbles and disk their paddocks (i.e., leaving no organic matter on the surface), thus increasing the risk of drought-induced soil erosion.

Secondly, irrigation, perhaps the most obvious technological solution to the agricultural drought problem, has only limited potential application in Australia. Even when applied, it has not provided 'drought-proofing.' Currently, only about one-third of a percent of the total pastoral and agricultural properties and only 9% of the total cropped areas are irrigated. For some crops, irrigation is admittedly vital. One such crop is cotton, but in the 1982–3 drought cotton farmers in north-central New South Wales received only 10% of their pre-drought water allocation, and that was 30% greater than the pre-drought price.[5] For drought overall, technological responses have been less than adequate.[6]

Table 3. The economic impact of the 1982–3 drought in Australia

1. 67% of the national pastoral and agricultural properties were officially recognized to be drought affected (i.e., 67 000 properties)
2. Wheat crop failures led to an average of a 45% decrease in the quantity of wheat sold per farm and wheat receipts per farm declined 58%
3. The 1982–3 wheat harvest (7.8 million metric tons) was 47.6% of that for 1981–2 and the average yield (0.76 metric tons per hectare) was 40% down from the previous 5-year average
4. National employment reduced by 2% (i.e., 100 000 persons)
5. The overall rural output declined by 18% leading to a 1.1% decline in national output
6. Industrial effects were estimated to be as follows:
   Chemical fertilizer production for 1982–3 fell by 11%
   Flour and cereal production for 1982–3 fell by 10%
   Railway transport receipts for 1982–3 fell by 6%
7. Commonwealth and State Governments paid out disaster relief as follows (in Australian dollars):

Loans:	1980–1	=$60 million
	1982–3	=120 million
Freight subsidy:	1982	= 47 million
Fodder subsidy:	1982	=100 million

Sources: Estimates by Commonwealth Bureau of Agricultural Economics and by Campbell et al.[2]

Using the model from research on natural hazards, we might ask first, 'Have changes in human activity increased the exposure to environmental hazards such as drought?'; secondly, 'Have changes in the environment increased the chance of hazards such as drought being experienced?'; and thirdly, 'Have the reports of drought occurrence always been valid?' A brief review of examples from southeast Australia provides evidence relevant to these questions.

Changing human activity

For many of the closely settled agricultural areas in southeast Australia, European land settlement from 1788 onward took the form of a sequence with, first, pastoral land use and then cereal grain farming in the drier areas and, later, dairying in the moister areas. This sequence itself increased the vulnerability of the community to drought, since the rainfall needs of the pastoralist on the open range were generally less than those of the grain farmer or dairyman. The sequence implied both spatial shifts and intensification of activities at specific locations; both had further implications for vulnerability to drought.

Toward a lower and more variable rainfall country

The historical geography of crop and livestock production in Australia has been the story of an advance from the humid southern and eastern coasts toward the arid interior of the continent. This advance toward the drier country, however, was not merely down the gradient of decreasing mean annual rainfall, it was also into areas of greater variability of seasonal rainfall. Not only were the absolute totals of mean annual rainfall declining inland, but the chances of obtaining that rainfall were similarly declining.

In South Australia the expansion of wheat farming in the 1870s from the relatively safe Mid-North to the more risky Willochra Plains has been well documented.[7] Popular pressure to turn pastoral lands into farm lands brought a temporary influx of farmers, who then suffered heavy losses when a prior brief moist season sequence reverted to the more usual semiarid conditions (Figure 23). A similar expansion into the northern Murray Mallee took place in the 1920s and 1930s with similar results.[8]

By the 1880s the farmers had advanced from areas where the frequency of droughts sufficient to prevent crops maturing was only two to three years in ten, to land where the frequency was six years in ten; in the 1920s and 1930s farmers left areas where four years in ten were drought, for lands where droughts occurred in more than six years in ten.[9] Government drought relief was generous but too short-term and insufficient to offset the toll of increasingly frequent drought losses. Only a major restructuring in both areas (as part of the officially designated 'Marginal Lands' scheme by which bankrupted or failing farmers were bought out

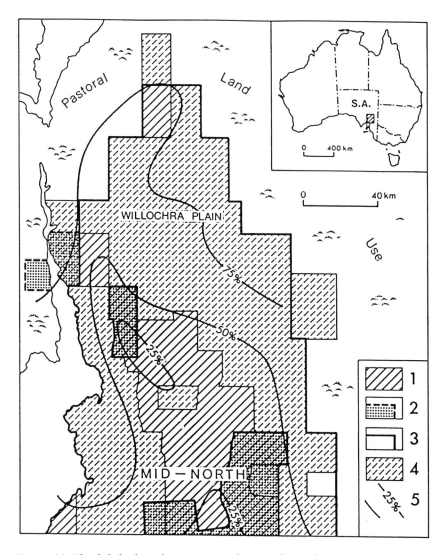

Figure 23. The failed wheat frontier in South Australia in the 1880s. 1, Districts (hundreds) with wheat yields of 5 bushels per acre and above 1880–2. 2, Districts (hundreds) open for agricultural settlement by 1869. 3, Districts (hundreds) open for agricultural settlement by 1879. 4, Districts (hundreds) with wheat yields less than 5 bushels per acre 1880–2. 5, Isolines of drought frequency (percentage of years with inadequate rainfall in the wheat growing season; from long-term records).

and their lands reallocated), along with pressure to introduce mixed grain and livestock farming (and so reduce the risk of the adverse consequences of drought), was able to maintain a thinned settlement during the 1930s and 1940s. Both areas remain among the most drought-prone grain-producing areas in the State, and their future as grain producers is uncertain.[10]

In New South Wales the spread of wheat farming into the interior brought farmers from the coast, where the grain suffered from fungal diseases induced by excess moisture but where rainfall variability (average absolute deviation from normal expressed as a percentage of normal rainfall) was only 20%, to, in 1901, areas where the variability was 25–30% and, by the 1980s, into western areas where variability was 30–35% (Figure 24). This westward march of the wheat frontier was aided by the breeding of drought-resistant wheat varieties and new farming technologies, and was the source of much official pride, as the farmers approached and locally passed the 254-mm (10-inch) isohyet of mean growing season rainfall. Yet, in times of drought it is these western wheatlands that suffer most, with more variable yields and a higher proportion of variation in yield explained by rainfall.[11]

Intensifying the pressures on the land

The severity of drought impacts reflects in part the demand placed upon the environment by human activity. At one level of activity rainfall may be adequate but for an intensified level of activity – say a different land use from what we have seen – the same rainfall may be inadequate. Mismanagement of the resource or economic necessity may also precipitate drought conditions.

Overstocking the range, whereby, for example, perennial shrubs are eaten out as well as the annual grasses and herbs, inevitably means that future drought impacts (as manifested in stock losses) will be more severe, since there will be no reserve of perennial plants, and inedible plants may well have taken over the niches of the destroyed edible species. One of the reasons given for the major losses in eastern Australia during the droughts of 1895–1902 (losses of half the sheep flocks and about one-third of the cattle

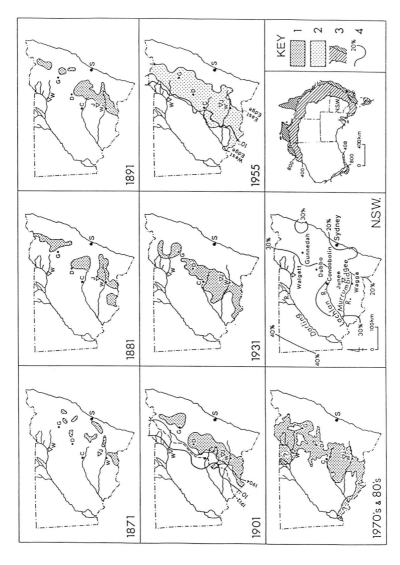

Figure 24. The advance of the wheat farmers in New South Wales, 1871–1980s. 1, Areas of wheat production at dates indicated. 2, General area of wheat production in 1955. 3, Median annual rainfalls for Australia, in millimeters. 4, Isolines of rainfall variability for New South Wales.

herds) was that pastoralists had overestimated the carrying capacity of the range and the resultant overstocking had been 'eating the haystack.'[12]

Economic pressures on pastoralists and farmers may require management strategies or create conditions where vulnerability to drought is enhanced. The most obvious situation was where the resource manager was required to intensify production from the same area because the product price was declining relative to costs of production or cost-of-living indices. This was a basic factor exacerbating the situation on the marginal wheat lands in the 1930s, when depressed wheat prices forced farmers to sow more area only to lose more when the droughts struck again. It is still an important factor. A 1970s study showed that in western New South Wales 51% of all pastoral properties had an officially assessed stock carrying capacity less than that thought to be economically viable, and a further 26% were borderline cases.[13]

The pressure on such property managers to 'overstock' in order to remain economically viable must be considerable, even though they are aware that by being overstocked they are more vulnerable to drought. Significantly, one official response to the dilemma has been to allow such pastoralists, normally not permitted by their lease to grow crops, to sow areas of their range to wheat in the hope of obtaining additional income. Such so-called opportunity cropping is allowed after above-average rainfalls and has even been practiced on the drying-out margins of ephemeral lakes with some success. This successful cropping, however, is transitory, and such areas were among the worst affected by wind erosion during the 1982–3 drought in New South Wales, as the cultivated areas provided neither crop, livestock feed, nor protection against soil movement (Figure 25). Again, the vulnerability to drought was increased.

On the South Australian cereal grain farms, over the years, the economic break-even point in wheat yields has been increasing. In the 1880s an average of 0.2 metric tons per hectare was needed to make a profit; in the 1930s it was about 0.4 metric tons per hectare; in the 1950s and 1960s it was about 0.6 metric tons per hectare; currently, it is thought to be about 0.8 metric tons per

Figure 25. Cover of the *Journal of Soil Conservation*, New South Wales, July 1983, showing 'Dust storm at Condobolin, January 1983' at the height of the 1982–3 drought.

hectare and is still rising.[14] On this basis the same rainfall which gave a bonanza wheat crop (say, 0.7 metric tons per hectare) in the 1880s would be classed as a drought in the 1980s! Needless to say, the areas where yields are most likely to fall below the break-even point are at the inland arid edge of the wheat areas and include the Wollochra Plain, northern Mallee areas, and Eyre Peninsula.[15]

Changing environment?

Interseasonal variation in rainfall and its relationship to human activity may be only one factor affecting the incidence of drought. The environment of southeast Australia has also changed over the years since European settlement began in 1788, and some of those changes have influenced the probability of drought occurrences.

Part of the explanation for the enthusiastic spread of grain farming into the semiarid interior in the latter half of the nineteenth century has been based on the suggestion that the climate of the period 1881–1910 was on average moister than that of the following 30 years.[16] The revival of such expansion in the 1960s and 1970s in New South Wales may reflect the claim that the climate of the period 1946–74 was moister than the previous 31 years.[17] The fact that the Commonwealth drought relief and marginal land schemes came into operation to rescue debt and drought-ridden farmers in the southeastern states and Western Australia during the intervening drier period in the 1930s and 1940s[18] seems, with hindsight, to have been an inevitable response to this apparent environmental variation.

However, it is not only the climate that has been quite variable. The environmental impact of settlement in Australia since 1788 has been considerable – over two-thirds of the continental vegetation cover has been modified by thinning, grazing, or complete removal.[19] Quaternary dune systems stabilized under indigenous scrublands have been reactivated, when droughts killed off the wheat in the paddocks cleared of scrub. In the drought of 1983 dust from overgrazed ranges and bare croplands fell on the city of

Melbourne, just as it had in the drought of 1902. Years of mechanical tillage have created impervious 'plow-pans' in some soils such that neither moisture nor crop roots can penetrate below the shallow surface zone and the vulnerability to drought stress is increased.[20] The undoubted benefits of the spread of settlement have clearly had their associated costs.

How accurate has drought reporting been?

So far, we have assumed that reports of drought occurrence have been valid and reasonably accurate. Evidence exists, however, that sectors of the Australian rural community have been 'crying wolf' on drought for some time.[21] Accusations of illegal use of official drought relief funds in the 1980s in Queensland led the Commonwealth government to remove drought from the list of natural disasters for which national relief funds were available to the States. At the same time, it set up an inquiry into future drought relief policies. That inquiry suggested in its final report that

> [agricultural] drought is a relative concept ... It reflects the fact that current agricultural production is out of equilibrium with prevailing seasonal conditions. This may be due to the degree of climatic variability and/or to inappropriate or inflexible levels of agricultural activity.[22]

Drought was not to be reinstated as a natural disaster but improved on-farm management was to be encouraged instead.

Why had this misuse of national funds occurred? There may be many reasons, ranging from ignorance of the aims of the relief policies to selfish greed, but part may rest in the historical trend of declining profitability of the rural sector in Australian agriculture. Not only did the climatic variability increase over time as the farmers and graziers pressed inland, but economic variability increased as well. With a tightening cost-price squeeze, the pressure on the land increased and, not only did drought frequency appear to increase, but the attraction of drought relief payments as a survival strategy increased also. It paid to 'cry wolf.'

Summary

Does the history of land settlement in Australia since 1788 suggest that drought has followed the plow? A brief review such as this cannot hope to be conclusive but the advance of human activities down the rainfall gradient, the increasing pressures of the commercial production system, the transformation of the environment as part of the settlement processes, and the seductions of regular disaster relief have, often imperceptibly, increased the vulnerability of the community to drought. The community has, of course, defended itself through its innovative agrotechnology and water management systems. Yet, one is left with the impression that drought in Australia may indeed have followed not only the plow, but the domestic livestock on their long march into the arid interior.

Ethiopia

JAMES McCANN

Because of its several major droughts and famines in the 1970s and 1980s, Ethiopia has joined the ranks of India, China, and the West African Sahel as yet another region of the world that has been plagued by chronic food shortages in the twentieth century. As a case study, it illustrates the causes, effects, and responses to drought and famine and the relationship between these two phenomena.[1]

Ethiopia's geography is dominated by its volcanic highland zones (Figure 26). The high plateaus range in elevation from frost-prone zones over 3000 meters at the highest point to the arid plains below the surrounding escarpment between 1000 and 1500 meters in elevation. Ethiopia's highland plateau is bisected by the Great Rift Valley, which creates northern and eastern plateaus separated by a system of deep Rift Valley lakes. On the highlands, sharp differences in elevation also define patterns of temperature, rainfall, and growing seasons. Most of Ethiopia's ox-plow agriculture has taken place historically in the elevation zones between 1500 and 3000 meters, where rich soils and a wide range of cultivars of cereals and pulses have sustained complex state systems for over two millennia. Rainfall ranges from below 500 mm per annum in the lowlands to over 2000 mm (Figure 27). Below the escarpments to the east and west of the highlands, annual rainfall and its reliability decline sharply.

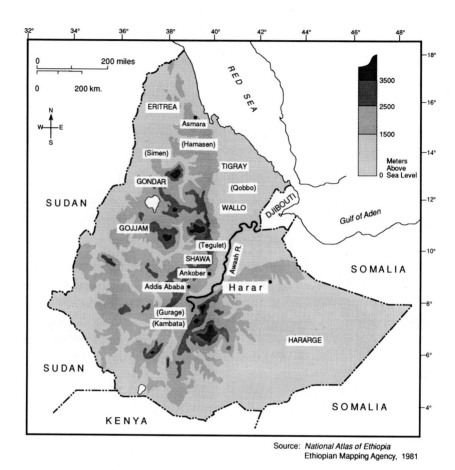

Source: *National Atlas of Ethiopia*
Ethiopian Mapping Agency, 1981

Figure 26. Ethiopia. Names in capital letters are provinces; names in parentheses are districts. (NB: The province of Eritrea became an independent country in May, 1993.) After J.C. McCann, *People of the Plow: A History of Agriculture in Ethiopia, 1800–1900*. Madison: University of Wisconsin Press, forthcoming.

The areas of Ethiopia that are most adversely affected by drought are generally those lands with highly variable interannual rainfall considered marginal for sustained agricultural production. For Ethiopia, marginal lands include newly cultivated areas as well as land historically productive but whose fertility has increasingly declined. There is also a strong correlation between famine-prone zones and areas that rely on two growing seasons a year: the unre-

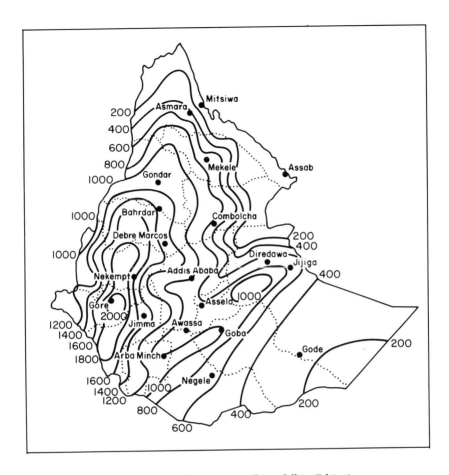

Figure 27. Map of mean annual rainfall in Ethiopia.

liable small rains in the spring (*belg*) and the main rains in the summer (*meher*). *Belg* rains play a critical role in shaping the rural economy of famine-prone areas.

Paradoxically, those areas of Ethiopia most affected by food shortages are in the central and northern highlands and have the longest history of sustained settled agriculture, and of Africa's most efficient labor farming system. The important issue with respect to climate is not that Ethiopia's climate may have changed during the twentieth century, but that human activities have changed the environmental and societal impacts of meteorological drought.

The level of impact of these tragedies has not only been determined by natural shocks to the rural economy, but has been affected far more by the makeup of the regional and household-level economies of the at-risk population. Even if one assumes that the frequency or intensity of environmental shocks has remained constant, human response factors have changed.

The historical distribution, density, and growth and decline patterns of the rural population have had a major effect on agricultural productivity and, in turn, on the vulnerability of the human population to famine in highland Ethiopia. Existing evidence strongly suggests that there have been substantial variations in population density in the northern highlands in both space and time during the twentieth century. Indeed, the rates of growth and decline in population historically account for a great deal of the variation in the forms of agriculture, uses of labor, and land tenure arrangements. In a number of areas, the relatively low ratio of population to land and natural resources allowed for economic expansion, while in most areas population pressure suppressed both the gross product and the per capita distribution of resources. The result has been a cycle of the progressive 'squeezing' of productive capacity within the agricultural economy in a north-to-south and east-to-west pattern. The long-term reduction of output by region and by farm has meant decreasing access to sufficient resources necessary for agricultural production within local rural economies.[2]

To this core area of vulnerability of the central and northern highlands must be added new areas – parts of Gurage and Kambata to the south in Shawa and highland Hararge to the east – where population densities have grown rapidly in this century and either equal or exceed those of the northern areas of longest settlement. Finally, newly inhabited marginal lowland zones along the eastern escarpment have joined the famine-vulnerable zones, primarily because they have received highland migrants during the past few generations. These lowland zones have highly variable rainfall seasonally and between years, and the combination of in-migration with a high level of climate variability has created a new famine-prone zone.

The notion that drought follows the plow can be effectively applied to Ethiopia for the following reasons: (1) agricultural activities have moved into areas where rainfall from one year to the next is 'normally' erratic; and (2) farm productivity in the older cultivated areas has declined to a point where even small variations in annual rainfall have undermined the viability of 'normal' agricultural production activities.[3] This section examines some areas of historical change which have contributed to the movement of smallholder agriculture into marginal lands and the consequent increase in the vulnerability of agricultural activities to meteorological drought.

Demography: shifting vulnerability

Historical records for the first two decades of the twentieth century indicate that major disasters that directly or indirectly affected human populations were concentrated in the northernmost regions of Eritrea, Tigray, and northern Wallo – particularly in their eastern sections. Disasters included locust infestations (1905–7, 1911–12, 1926–31), influenza (1917–19), and major drought (1914–16). The records also show that Eritrea and probably eastern Tigray were net importers of food and livestock at least since 1915, and most likely had been so since the middle of the nineteenth century. Northern Wallo probably followed shortly thereafter (after 1915) as a food deficit area.[4] During the 1930s, areas in crisis expanded southward, encompassing most of Wallo, and the northern and eastern sections of Gondar region. After World War II, southern Wallo joined the list of famine-prone areas.

By the 1980s, famine-prone areas had expanded to encompass a wide section of the central highlands, with northern Shawa and parts of highland Hararge being added to the crisis zone. The entire lowland belt running north to south below the eastern escarpment, an area largely unpopulated before 1920, has also become vulnerable to drought-related food shortages and famine. This increasing propensity toward vulnerability is a reflection less of changes in local or regional climate than of changes in demographic trends. A clear pattern of general population growth throughout the

1920s and into the post-1941 years emerges from the available records and is a critical variable in the geographical expansion of vulnerability.[5]

Population growth rates, migration, declining soil fertility and agricultural productivity, and the opening up of new marginal lands to agriculture are closely linked. By the early 1980s Tigrinya-speaking migrants from Hamasen had already arrived in the Wag region of Wallo.[6] From that period at least, Tigrinya-speaking people had penetrated southward into northern Wallo as farmers, but even more prominently as part of a mercantile migration which penetrated south along the caravan route to Shawa. In Hararge, cash income from the narcotic cash crop *chat* and coffee in other areas and the consequent relatively heavy involvement in a regional cash market economy have buffered the effects of the demographic explosion. However, the array of resources available to households in the form of animal traction, forage, and productive land had created a dependence on food-for-work grain payments in the late 1980s.

In the Simen region, aerial surveys indicated that the number of household dwellings (farm buildings) rose at an annual rate of 2.4% during the period 1955–75, in the most populous and productive elevations. By contrast, the number of rural dwellings in the lowlands rose by 3.58%, a measure of migration from adjacent highland zones. Since 1964, the upper (altitudinal) level of cultivation in the Simen region has climbed 100 meters to just below the frost-line limit.[7] Such evidence from a region which historically and ecologically resembles areas of western Tigray, western Wallo, and parts of Gojjam strongly suggests that migration and the expansion of cultivation are a product of population growth and a population's overconcentration on agriculturally productive land.

The highland Tegulet district of northern Shawa to the west of the escarpment closely resembles the progression over time of the population growth evident in the Simen study. In that district population pressure has mounted steadily since at least the end of World War II, but probably began much earlier. In Tegulet the specter neither of drought nor of famine has been a regular part of the experience of the rural population. Yet, under the pressure of population growth, pastures and fallow land have virtually disappeared, and the average size of landholdings has declined to less than 2 hectares. Even the river floodplains, 1000 meters below the plateau, have been put into cultivation within the last generation.[8]

This pattern parallels the much broader historical trend of emigration to less productive and more risky (with respect to agricultural output) lowland cultivation evident along the eastern escarpment. The entire lowland belt running north to south below the eastern escarpment, an area largely unpopulated before 1920, has also become at higher risk to drought-related famine. This progression is a reflection less of changes in climate than of demographic trends. Below the Addis Ababa to Asmara road (the former caravan route), areas such as Qobbo in Wallo or Denki in northern Shawa, lowland zones which had primarily been pastoral wet-season grazing areas, have been put into cultivation in the past generation or two by migrant highlanders. In the lowland regions east of the old Shawan capital of Ankober, this pattern of regional population dynamics has emerged primarily in the past two generations.

The Ankober region's history reveals the effects of variations in demographic pressure and productivity. The region as a whole has received a steady influx of Amhara settlers since the mid-nineteenth century.[9] In the highland regions general patterns of ambilineal (male and female) descent have changed in response to the threat of further fragmentation of farm land. Prior to 1974, elder siblings were able to 'buy off' the younger family members' land claims in an attempt to preserve the economic viability of their already small holdings. Life histories of present inhabitants reveal a consistent movement into these lowland areas by Christian farmers who were the younger, dispossessed children of households located in the highlands to the west. New migrants arrived first as tenants, using sharecropping agreements with local Muslim landholders. They have continued to arrive, since the 1975 land reform, to claim through peasant associations the diminishing stock of open land. Household holdings in these lowland areas average 2 to 2.5 hectares, substantially above the adjacent highland average of 1 to 1.5 hectares.[10]

The Denki district (5 hours by foot northeast of Ankober) is a microcosm of the lowland belt which has a low agricultural potential resulting from ecological conditions (e.g., it is drought-prone, has limited dry season pasture, and endemic human diseases) rather than poor soil fertility or absolute population pressure. Overall, the increase of population in such areas, as a result of in-migration as well as natural population growth, has substantially increased the proportion of the highland population that is susceptible to the consequences of adverse climatic variations. Low agricultural productivity has, thus, been the result less of land shortage than of problems associated with climate variability inherent in lowland zones where meteorological drought, particularly during the regionally important *belg* season, is part of the local climate and plays such a critical role for marginal farms.[11]

In addition to the loss of crops, the greatest effect of drought has been to deplete the supply of oxen, many more of which die from the lack of forage than the lack of water. Thus, the population faces a double jeopardy: the high variability of the rains makes farmers vulnerable to the impacts of drought, while the low level of the natural resource base has placed much of the population in danger of drought-related food shortages leading more quickly to famine.

The spatial expansion of cultivation into areas marginal to agricultural production has also reduced the ability of pastoral economies to adjust their transhumant patterns to climatic variability. For example, Afar pastoralists who have traditionally moved between pastures in the Awash valley and the middle altitude zones along the eastern escarpment have, in the past two decades, lost their dry season pasture to irrigated agriculture along the Awash River and to subsistence agriculture in the eastern lowlands. In some cases they have even been settled in marginal agricultural areas.[12]

Technology, social institutions, and vulnerability

An important dimension of vulnerability in Ethiopia's rural economy has been the failure of smallholders to adopt over the past century innovation in design, materials, or application of technology (irrigation, mechanization, agronomic techniques). The basic ox-drawn, single-tine plow has remained the basis of production throughout the highlands. Surprisingly, the shift from oxen to horses, or from scratch to moldboard plows, or to a three-field rotational system which characterized European adaptation to population pressure has not taken place broadly across the highland farming system.[13] In some cases adaptive forms of land use such as terracing and irrigation have been abandoned rather than expanded in the face of population pressure. The mechanization of smallholder agriculture, even in the most developed zones, has been marginal at best.[14]

Ethiopian historian Merid Wolde Aregay has blamed technological stagnation on the social system inherent in peasant land tenure and the insecurity of income rights of the elite in the agrarian political economy. He has argued that the system of land tenure and partible (divisible) property rights (*rist*) together with elite control of income rights (*gult*) combined to stultify technological initiative. As Merid suggested, examples of new forms of technology abound, but their systematic application has been extremely limited.[15] Indeed, the fragmentation of political authority which characterizes highland political culture has inhibited the cooperation necessary to sustain irrigation, especially in the face of demands on upstream resources resulting from increasing demands of a growing population. Variability in climate and crises in food production, far from stimulating innovation, appear to have driven smallholders further toward conservative risk-averse cropping strategies.

The impasse of technology amid crises of diminishing resources in land and capital has placed a severe limit on the agricultural economy's ability to increase crop yields as well as total food production. The traditional equation involving land/labor/demographics has depended on specific technological boundaries. The failure to alter those constraints over time has confined productivity to the limitations of local and regional agricultural cycles of population growth and decline.

The effect of slow technological change has been a steady decrease in gross profit per farm and an expansion of vulnerability with the reduction of farm size. At the same time the number of oxen per farm appears to be declining along with farm size. Thus, the number of plowing hours has been declining steadily since the

1920s, an excellent measure of expanding structural vulnerability and the impoverization of rural producers.

Social relationships between and within rural households have contributed to vulnerability. An enduring principle of rural household relations in the northern famine-prone regions has been the weakness of horizontal ties between households and the relative strength of patterns of dependence which link households of unequal resources. Traditionally, these inequalities have generated an elaborate network of debt and dependency which deepen during hard times but relax during economic recovery. There have also been clear patterns of vulnerability within households which derive from the particular nature of intra-household relations based on gender, age, and status. Although drought may bring this vulnerability into sharper relief, the weakness is already a structural part of social relations set by preconditions of social conventions of property, marriage, and inheritance and stages in the household development cycle.[16]

Expanded vulnerability to drought has been the result not only of the movement onto marginal lands but also of the characteristics of households which have made that move. It is not just that drought has made farmers poor but that poor farmers have moved into areas with higher risk of drought. Ironically, as land resources have become scarce relative to increasing population, capital resources as much as (or more than) land have become scarce factors in production and, therefore, yet another source of vulnerability. For the most part it has been the undercapitalized farmers who have shifted to the lower (i.e., less productive) elevations. Consequently, social relationships for the distribution of labor, capital, and debt have become increasingly important for determining which members of society are most vulnerable to climate-related shocks.

In most northern highland areas local institutions for the borrowing and transfer of oxen and seed have been central to establishing local patterns of debt and dependency. *Belg* crops in these areas have become particularly critical on smaller, resource-poor farms which require capital and credit for obtaining food in the 'hungry season,' that is, the period just before the main season's harvest in December and January. When the *belg* rains fail, poor farmers must borrow new seed, obtain food, and pay higher prices for obtaining oxen in order to prepare the land for planting at the beginning of the main rainy season. Capital resources such as livestock and seed are among the first to disappear in the face of a meteorological (or economic) shock. The social institutions which link wealthy farmers to poor ones through accumulated debt have tended to condition the nature of vulnerability. The relationships of debt developed through the borrowing of animal traction, seed, and food to the reduction of overall farm productivity and gross output are precisely those which emerge during famine conditions.

The relative vulnerability of household members depends to a large degree on their gender, age, and social status and consequent access to productive resources. Thus, households in the northern Ethiopian tradition are neither transgenerational property estates nor extended family units. Households dissolve on a regular basis. There is a high rate of divorce, while death and economic insolvency also account for household dissolution. The welfare of a given household over time is not necessarily a reliable indicator of the welfare of its constituent members.

Field work on this issue has indicated that, as a result of divorce, women in particular have been subject to a loss in their access to oxen and other household capital assets, and particularly to the loss of male labor. Evidence from field work also suggests that female offspring appear to be less likely than their male siblings to inherit capital goods and to receive them through marriage endowments. At the same time the relative weakness of women, the elderly, and the very young in access to property, agricultural labor, and claims on other resources emerges only during crises or the dissolution of the household.

In 1975 the Provisional Military Government of Ethiopia promulgated a land reform program which addressed some of the questions of women's rights to land and recognized their problems in access to labor.[17] Unfortunately, land reform has not effectively altered patterns of property (access to oxen, tools), normative gender roles in labor, or marriage law. In fact, overall, agrarian reforms have increased the demographic pressure on productivity rather than reducing it.[18] In most areas land reform has increased demand for allocations of land as a result of in-migration and natural population growth, resulting in an increasing fragmentation of landholdings and decreasing plot size. Some peasant associations have responded by limiting new membership and slowing allocations to young households. Moreover, the social institutions governing access to capital have not been altered, except in the small numbers of producers' cooperatives where capital is held jointly. Thus, the relative scarcity of key resources such as oxen, forage for livestock, and seed for planting has continued to provide the basis for rural stratification and structural vulnerability.

Several key characteristics of regions vulnerable to disaster are clear, given historical shifts in the nature of Ethiopia's rural economy. The critical issues are, in fact, corollaries of the more fundamental decline in agricultural productivity at the level of the smallholder farm. In sum, the movement to marginal drought-prone land has taken place as a direct result of decreasing farm size, decreasing each farm family's access to animal traction and credit, and the historical failure of small-farm technology and agronomy to intensify production. These, in turn, have prompted the movement of new population into zones with relatively higher rainfall variability and poorer soils, and have exposed the inherent underlying vulnerability of certain members of rural society whose weaknesses emerge only during shocks but not under general conditions of poverty.

In the past decade changes in the national political economy as well as in the international political system have also introduced new features of the vulnerability of farmers to disasters, such as the lack of access to employment, the lack of aid infrastructure (storage points, roads, security), the lack of investment in famine-prone rural areas, the lack of political security, and the lack of security of tenure over land and farm resources such as perennial crops and forests.

The heightened scope and intensity of Ethiopia's recent vulnerability to famine and natural disasters are in part an issue of climate variability and environmental change. More fundamentally, however, that vulnerability has been a function of changing societal factors and of a long-term decline of the small-farm economy. The size and composition of the rural population that has become susceptible to adversities during an environmental crisis has grown dramatically in the twentieth century, as a result of declining smallholder productivity. Farmers with access to less land, fewer oxen, and moving into areas subject to relatively higher rainfall variability may still engage in production, but their lack of access to employment, credit, and resources needed for intensification has given them an increasingly smaller margin of safety from climate-related shocks, not only in the agriculturally marginal areas, but in the higher-rainfall regions as well.

Northwest Africa

WILL SWEARINGEN

'Northwest Africa,' as used here, refers to Morocco, Algeria, and Tunisia (Figure 28). These are the North African countries that are dominated by the Atlas Mountains and, because of the orographic effects of these mountains, have higher precipitation and can support rain-fed agriculture. Virtually all of this region was controlled by France during the colonial era. As part of this legacy, the French still refer to it as 'Afrique du Nord.' Another common toponym for the region in both France and the United Kingdom is the 'Maghreb' – from the Arabic expression for the region meaning 'the West.' However, 'Maghreb' was not used for this study for two reasons: first, it is unfamiliar in most of the English-speaking world, and second, as commonly used, this regional term also includes the desert countries of Libya and Mauritania.

This case study examines the degree to which Northwest African societies are becoming more vulnerable to drought. Since 1980, Northwest Africa has suffered frequent and severe episodes of drought, raising fears of climate change. While not denying that climate change (e.g., global warming) may be occurring, it is argued here that the risk of drought impacts has been increasing primarily for socioeconomic reasons. Key factors include rapid population growth, the effects of French colonization, agricultural practices, environmental constraints, and government policies.

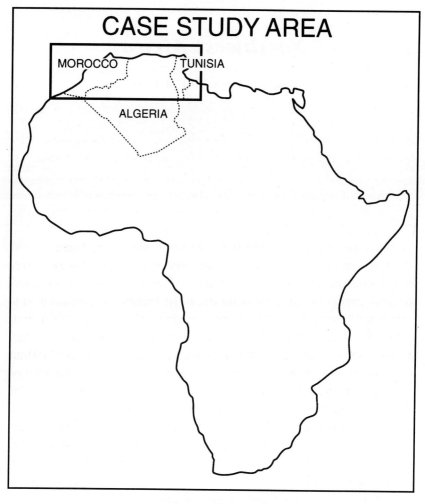

Figure 28. Northwest Africa.

The environmental and social context of drought

Northwest Africa is naturally vulnerable to drought. This region is situated on the southern margins of mid-latitude weather systems arriving from the northern Atlantic. As a result, both the timing and total amounts of rainfall within a given year, and between years, are extremely erratic. Climatic variability introduces a high degree of risk in rain-fed (non-irrigated) agriculture. Furthermore, precipitation levels are generally insufficient for sustained rain-fed agriculture in most of the region.

Drought is the region's dominant natural hazard and occurs frequently in all three countries. Morocco's drought history is representative: During the present century, Morocco has experienced approximately 25 years of agricultural drought. Each of these Northwest African countries experiences roughly the same frequency of drought. However, drought in one country is often not correlated with drought in the other two countries. For example, in 1988 Morocco had the largest cereal harvest in its entire history, while Tunisia suffered its worst harvest in over 40 years from drought-related crop failure.

The profound socioeconomic significance of drought in Northwest Africa results from the crucial importance of rain-fed cereal crops in farming systems and diets in the region. Since classical antiquity, Northwest African farmers have specialized in cereal cultivation (Figure 29). Wheat and barley are the primary cereal crops; however, maize and oats are important secondary cereals. In Roman times, and during much of the French colonial period (1830–1962), Northwest Africa was a major exporter of grains to Europe. Today, cereal crops still account for approximately 85% of the region's cropland, and are primarily produced by rain-fed agricultural methods. Wheat, consumed both as bread and *couscous*, is the mainstay of the national diets in the region. Nutritional surveys reveal that cereals account for nearly two-thirds of the caloric intake and protein in the average Northwest African diet. The percentage is even higher for the urban poor.

Droughts in Northwest Africa diminish both cereal hectarage and yields. The resulting declines in total production pose an immediate threat to food security. Typically, during a drought year, food shortages develop, food prices rise, hunger becomes more pervasive, peasants abandon their land and migrate to the cities, cereal imports rise dramatically, foreign debt increases, and the affected country's economy suffers a recession. For example, largely because of drought in 1986, Tunisia's Gross Domestic Product shrank by 1%, down from the previous year's 4.6% growth rate.

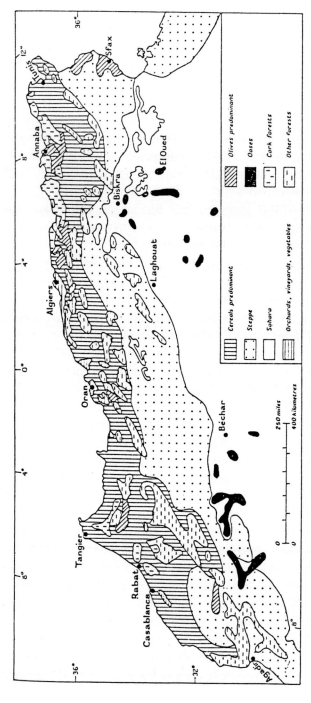

Figure 29. Agricultural zones of the Maghreb. From Harrison-Church et al., 1964: *Africa and the Islands*. Harlow, Essex, UK: Longman Group Ltd, 120.

Good harvests in Northwest Africa require at least adequate rainfall during both the October–December planting season and the January–April growing season. Conversely, disastrous harvests can result from shortages of rain during either season. Given the potential for extreme variability in precipitation, the 400-mm average annual precipitation line (isohyet) is normally considered to be the threshold for viable rain-fed cereal production in this region (Figure 30). However, the timing of rainfall is as critical as the total accumulation. For example, if the entire November precipitation in Tunisia falls during a single day, most of it will be lost to runoff. Thus, regardless of the total amount of rain, drought conditions will probably develop. By contrast, as little as 250 mm of annual precipitation can produce good harvests if rainfall occurs during the periods of greatest plant need.[1]

Farming technology plays a key role in determining vulnerability to drought in Northwest Africa. Somewhat over half the total tonnage of cereals is still produced with traditional technology, including animal traction and the light-weight swing plow (Figure 31). Approximately 75% of Morocco's cereal production is still by traditional means, compared with around 45% in Algeria, with Tunisia's percentage somewhere in between. Traditional technology helps to prevent soil erosion. However, it has the following critical shortcoming: farmers cannot begin to plow until the soil has been softened by approximately 150 mm of rainfall. This is because, during the summer dry season, the soil becomes too hard for the light plow to penetrate.

Having to wait for rain presents major problems. The window of opportunity for planting is relatively limited, given the short growing season before the onset of summer's hot, desiccating conditions. Thus, planting must generally be finished by late December or early January. If the autumn rainy season is delayed or rains are poor, the area of cropland that can be put into production is reduced.

The resulting decrease in planted area can be quite substantial. For example, during the 1980s the annual hectarage figures for cereal crops fluctuated by well over a million hectares in each country. The magnitude of these fluctuations was equivalent to slightly over one-half of Algeria's average cereal hectarage during

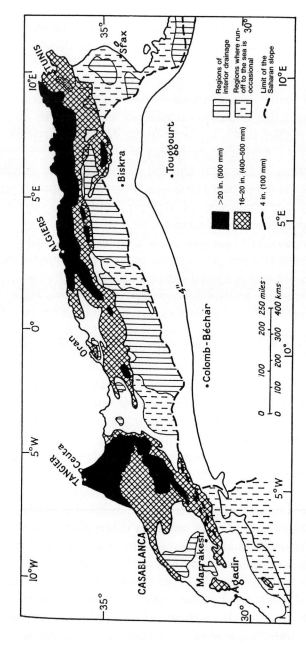

Figure 30. Rainfall and runoff in the Maghreb. From Harrison-Church *et al.*, 1964: *Africa and the Islands*. Harlow, Essex, UK: Longman Group Ltd, 107.

Figure 31. Traditional plowing in Northwest Africa.

this decade, and nearly one-fourth of Morocco's. In Tunisia, cereal hectarage in 1988 was down by 60% from the 1980–7 average because of lack of rainfall during the autumn planting season. These enormous fluctuations are principally the result of rainfall variations and the technological limitations of the traditional plow *vis-à-vis* Northwest Africa's physical environment.

Having to wait for rain before plowing not only reduces crop hectarage, it also significantly lowers crop yields. This is because a large percentage of the total precipitation has already evaporated before the seeds are even planted, given the slow pace of plowing with animal traction. The significance of this factor is demonstrated by field tests in Morocco, which have shown that planting with tractor and steel plow before the onset of autumn rains can boost crop yields by roughly 30% – without having to resort to any other changes in production methods.

Northwest African peasants are well aware of the constraints of traditional technology. Animal traction and the traditional plow have been partly replaced by the tractor and modern steel plow. All three Northwest African governments have actively promoted agricultural mechanization in recent years. The relatively more prosperous landowners have benefited from various government subsidies and have substantially mechanized their operations. Peasants with larger holdings who cannot afford to modernize often find it cheaper to hire plowing services than manual labor. This, too, has contributed to the expansion of mechanization.

However, the vast majority of the peasants in each country can afford neither agricultural machinery nor plowing services. Furthermore, their holdings are too small and fragmented for mechanized operations to be economically feasible. Ten hectares of dryland is generally considered to be too small a landholding to support a peasant family unless the land is irrigated; most are not. In Morocco, for example, more than 95% of the agricultural families have less than 10 hectares of land; in Algeria, 85%; and in Tunisia, 75%. Thus, animal traction and the traditional plow will continue to be used extensively. This will help to conserve the Northwest African environment. Unfortunately, it will also help to preserve the region's vulnerability to drought.

Increasing vulnerability to drought

An exhaustive historical survey of drought and other natural calamities in Tunisia has determined that there were at least 26 major drought-related famines during the period from 100 AD to the late nineteenth century. While drought has perhaps always been a major hazard for agriculture in Northwest Africa, the potential for agricultural drought has apparently been increasing as a result of processes initiated by colonization. Algeria became a French colony in 1830, Tunisia a French protectorate in 1881, and Morocco a French and Spanish protectorate in 1912. The colonial period lasted until 1956 in Tunisia and Morocco, and until 1962 in Algeria. In all three countries, colonization introduced a series of processes that have steadily increased the likelihood of drought.

Prior to the colonial period, agriculture in Northwest Africa consisted of an extensive system of rain-fed cereal agriculture and stockraising, with irrigated orchards and gardens surrounding most settlements. Oasis agriculture and pastoral nomadism were practiced in the desert regions. Most land was communally owned by tribal groups. Individual families within these groups farmed dispersed plots in an attempt to minimize the risk of crop failure. Surplus grain from bountiful harvests was stockpiled to cover crop failures during drought years. Fallowing was widely practiced. Fallowing both replenished soil moisture and helped to restore soil fertility. Low population density gave the arable expanses of primary interest to the French colonists a relatively underutilized appearance.

French colonial planners viewed Northwest Africa as a major exception within France's colonial realm. With its proximity to Europe, and the Mediterranean climate of its coastal plains and plateaus, the region was perceived to be conducive to large-scale French settlement. In all three countries, colonization dislodged peasants from much of the best land. Europeans acquired about 30% of Algeria's arable land (2.7 million hectares), nearly 20% of Tunisia's land (800 000 hectares), and 13% of Morocco's land (1 million hectares).

Exacerbating the effects of European colonization was land concentration by native large landowners. During the colonial period, indigenous landowners who allied with the French were able to amass sizeable landholdings in all three countries. In Algeria, about 25 000 native Algerians acquired a total of nearly 2.8 million hectares, more than 30% of the country's arable land. In Morocco, 7500 Moroccan landowners acquired 1.6 million hectares, or 21% of the arable total. And in Tunisia, 7200 Tunisians acquired 630 000 hectares, 15% of the arable land.

Land concentration had two profound consequences. First, peasants became concentrated on a diminished amount of arable land. Much usurpation was conducted by private land speculators, settlers, and the more prosperous indigenous landowners. However, French colonial authorities exercised a deliberate policy of *cantonnement*, that is, they calculated how much of a given tribe's territory was needed to support the tribe. They then 'cantoned' tribal members on that land and appropriated the rest for colonization. A major adverse effect of land expropriation was that it reduced the ability of peasants to let part of their land lie fallow. Reduction of the fallow area significantly increased the potential for agricultural drought. Fallowing substantially boosts the available water supply for subsequent crop use. Nearly a fourth of the precipitation falling during the fallow period (roughly 18 months between harvest and planting) is retained in the soil. In semiarid regions, this soil moisture component is often the critical difference between a successful harvest and a drought-related crop failure. As the proportion of land left fallow was gradually reduced, Northwest Africa's vulnerability to agricultural drought gradually increased.

Second, large masses of peasants were dislodged to marginal land that was not sufficiently attractive for colonization by the French. These marginal areas were commonly characterized by deficient rainfall, and possibly poor soils and excessively steep slope. Previously, most of this land had been used only for stockraising because of its poor agricultural potential. Once under cultivation, it was often highly vulnerable to meteorological drought.

Simultaneously, other significant changes were occurring during the colonial period. Public health measures introduced by the French caused native death rates to decline sharply. As a result, Northwest Africa's population began to expand dramatically, more than quadrupling during the colonial era. This population explosion – combined with the expropriation of from one-third to over half of the arable land by Europeans and indigenous large landowners – intensified pressure on remaining agricultural resources. Communally-owned land gradually became privatized, peasant holdings became increasingly fragmented and miniaturized through inheritance, the proportion of arable land left fallow continued to decline, additional rangeland was put under the plow, and increasing numbers of peasants could not be sustained on the land. By the 1930s, squatter settlements (known as *bidonvilles*) were developing around most major Northwest African cities, populated by refugees from the countryside.[2] Rural-to-urban migration was especially pronounced during drought years, when crop failures compelled many peasants to sell or abandon their land.

Colonial agricultural policy *per se* also played a major role in deepening Northwest Africa's vulnerability to drought. From 1915 to around 1928, colonial authorities in all three countries were ordered to promote cereal production for France. The architects of this mandate were convinced that France's 'Afrique du Nord' had been a bountiful breadbasket for Rome during classical times, and that France could restore this land to its former productivity.[3] High market prices and various subsidies and bonuses were offered to encourage cereal cultivation, particularly soft or bread wheat cultivation by mechanized means. These government incentives enabled marginal areas to be profitably cultivated. Although colonial farmers were the primary beneficiaries of government policy, this policy also encouraged Moroccan farmers to expand substantially the area planted in wheat.

Unfortunately, wheat, the primary crop promoted by government policy, was not particularly well suited to the lower-rainfall areas of Northwest Africa. Previously, the predominant native cereal had been barley – a crop better adapted to the region's environmental realities. With the varieties at the time, the critical soil-moisture needs for barley were about 30% lower than those for wheat. In addition, because barley ripened earlier than wheat, it could be harvested earlier and was less vulnerable to the untimely onset of hot, arid summer conditions.[4] Despite these considerations, colonial policy favored wheat production over barley, and wheat gradually became predominant. Since then, barley has unfortunately come to be viewed as an inferior food grain by many Northwest Africans. In short, by substituting wheat for barley, colonial policy increased the potential for agricultural drought.

From the perspective of the time, colonial policy was highly successful. In Tunisia, for example, the area planted in wheat increased from around 530 000 hectares during the period 1909–

14 to an average of 810 000 hectares in 1931–5, an increase of roughly 53%. In Morocco, the total cereal area expanded by nearly 60% between 1918 and 1929. Cereal cultivation in neighboring Algeria also expanded dramatically. However, there were hidden adverse effects. Part of the new cereal hectarage came from the reduction of fallow, increasing the potential for drought. Much of the rest came from the cultivation of marginal low-rainfall areas. Thus, the proportion of cropland vulnerable to drought steadily increased.

Agricultural policies since independence and the food security crisis

Since independence, each of the Northwest African countries has pursued a different development strategy. Algeria, emerging from a traumatic colonial experience and devastating war of independence, has attempted to achieve economic independence through a comprehensive program of industrialization. Morocco, by contrast, has emphasized export agriculture, investing heavily in the irrigated production of citrus and market vegetables. Tunisia has adopted the most balanced development strategy, investing in export agriculture and encouraging export-led industrialization by multinational firms.

All three countries recovered ownership of colonial landholdings and have engaged in limited land reform. Although a minor percentage of landless or land-poor peasants have benefited from land reform, much of the former colonial land has passed into the hands of rural notables or urban elites. Most of the large landholdings acquired by indigenous landowners during the colonial period were never subject to land reform.

For at least two decades following independence, the Northwest African countries seriously neglected domestic food production. As a result, by the 1980s all three countries were experiencing severe food security crises. The key symptoms of these crises were declining per capita cereal production, alarming and ever-growing levels of cereal imports, heavy foreign indebtedness related to these imports, and massive food subsidy programs.

By the early 1980s, Algeria was importing approximately two-thirds of its cereal supply, Tunisia was importing nearly half, and Morocco was importing over one-third. In each country a large percentage of the population was living precariously near the threshold of hunger. The political implications of this crisis became clear by 1981, when Morocco experienced a bloody food-related riot. Similar food-related riots erupted in Tunisia in 1983–4, in Morocco in 1984, and in Algeria in 1988.

Recent agricultural reforms

Since the early to mid-1980s, all three countries have been undertaking major agricultural reforms. The overriding objective has been to increase dryland cereal production. Reforms include the privatization of the state agricultural sectors to improve efficiency, as well as the promotion of modern seed varieties and fertilizers. In terms of increasing risk to drought, however, the most significant reforms involve changes in crop prices, promotion of agricultural mechanization, and a 'new lands' scheme in Algeria.

Since independence, Northwest African governments have maintained tight control over producer prices of basic food crops. Prices for these crops, cereals in particular, had been held artificially low until the 1980s. For much of this period, crop prices were no more than about half of what they would have been without government intervention.[5] The government's rationale was that low crop prices would enable them to provide cheap food to their urban populations, helping to keep wages low and thereby assisting industrialization and other urban development initiatives. An ulterior motive behind the cheap food strategy was to prevent social unrest among the growing ranks of the urban poor. Low crop prices, however, acted as a major disincentive to farmers, creating a vicious downward spiral of declining production.

Beginning in the late 1970s, fixed producer prices for cereals and other basic food crops were gradually raised. In Algeria and Tunisia, these prices approached world market levels by the mid-1980s. However, in neither country have higher prices yet increased production. Both cereal hectarage and total production levels remain substantially down from previous years. In Tunisia, for example, annual cereal hectarage during the 1980s was only 1.3 million hectares, compared with 2 million hectares during the 1950s – a decrease of 35%. In Algeria, this decrease was roughly 20%. Such reductions probably indicate a shrinkage of the natural resource base because of soil erosion and desertification, and possibly a period of poorer-than-normal rainfall.

In Morocco, changes in pricing policy have been far more dramatic. Here, the government doubled producer prices of barley and wheat, significantly exceeding world market levels. The stimulus effect was remarkable. Assisted by good weather, Moroccan production of cereals grew from a 1980–4 average of 3.8 million metric tons to 6.6 million metric tons during the period 1985–90. In part, this growth came from improved yields and in part from an expansion of the hectarage devoted to cereal production. Average annual cereal hectarage during 1980–4 was slightly over 4.4 million hectares. However, by 1990 it had expanded to 5.5 million hectares – an increase of 25%.

This increase came both from the reduction of fallow and the expansion of cereal cultivation into low-rainfall rangeland areas. Already by the early 1940s, cereal cultivation had reached the 4.4 million hectare figure that still prevailed during the early 1980s. In short, virtually all viable cropland had already been put into production half a century ago. The recent dramatic expansion of rain-fed cereal hectarage has been possible only because of a period of higher-than-normal rainfall. This has allowed fallow hectarage to be further reduced and the extension of cereal cultivation to rangelands that are normally too arid for rain-fed cereal crops. Encouraged by high crop prices, farmers have essentially been mining the fragile soils of marginal lands, lands prone to wind erosion, desertification, and drought.

Government efforts to promote mechanization have facilitated the extension of cereal cultivation to drought-prone rangelands. The tractor and disk plow have colonized large stretches of rangelands in all three countries. In southern Tunisia, for example, roads created by oil exploration crews have enabled mechanized farmers to penetrate regions that previously were accessible only to pastoral nomads. Similar penetration of previously remote grazing lands has also occurred in the other two countries. Some of the new lands being farmed normally receive as little as 200 mm of annual rainfall. Their poor soils can sustain cultivation for a few years, as long as higher-than-normal rainfall prevails. However, the return of normal (i.e., reduced) rainfall forces their abandonment and desertification processes quickly advance in the abandoned areas.

In Algeria, the cultivation of marginal lands has actually become official policy. In 1983, its government passed legislation that established an ambitious homesteading program, the overriding purpose of which was to encourage Algerian citizens to maximize the agricultural potential of the country through development of public domain land that had not previously been cultivated.

Prospective homesteaders are required to pay the nominal sum of 1 dinar (approximately US$0.12 in 1990), for which they are given an allotment of land within one of the numerous perimeters established by provincial authorities. The only stipulation is that they develop this land fully for agricultural purposes within a 5-year period, at the end of which they are granted unrestricted title, and are free to sell or use their land as they choose.

The government views the program as a way to expand the agricultural resource base, increase the food supply, combat peasant exodus to urban areas, and counterbalance excessive urban development along the country's northern coast. The goal is to put approximately 800 000 hectares of new land into production. About half of this land will be in the Saharan zone and involves small (less than 3 hectare) irrigated plots. However, the other half, some 400 000 hectares, involves larger dryland allotments in the country's high plateau region. Virtually all new 'cropland' in this region is low-rainfall steppeland suitable only for stockraising. The homesteading program, then, will significantly increase the proportion of Algeria's cropland in drought-prone areas. Thus, we shall hear more about droughts and crop failures in these areas in the future, but not because of climatic change.

The homesteading program, however, is only part of Algeria's current 'new lands' scheme. In 1984, the Algerian government initiated a comprehensive agricultural plan that includes the goal of putting 2 million hectares of new land into production. Two-fifths of this new land is to come from the homesteading program. The other three-fifths, or 1.2 million hectares, will come from the reduction of fallow in the traditional crop-rotation system. Such a major reduction of fallow substantially increases the risk (and, therefore, frequency and impacts) of drought in Algeria.

Drought does follow the plow in Northwest Africa

Drought is an endemic natural hazard in Northwest Africa. Even with proper land-use management, this region would be vulnerable to recurrent drought. Part of its vulnerability is associated with widespread use of the traditional plow. However, as this study has shown, the risk of drought has actually been increasing because of the following processes initiated during the colonial period: (1) the expansion of cereal cultivation onto drought-prone rangelands; and (2) the reduction of fallow. During the colonial period, these processes were fostered by large-scale land expropriation; by the dislodging of peasants to marginal lands; by a cereal policy offering high crop prices and other incentives; by agricultural mechanization, which facilitated the mining of marginal areas during periods of higher-than-normal rainfall; and by population pressure associated with rapid population growth. Other significant factors during this period included the gradual loss of the ability of peasants to stockpile grain as insurance against drought, and the progressive substitution of wheat for drought-resistant barley.

Since independence, populations in Morocco, Algeria and Tunisia have continued to increase at rapid rates. High population growth rates, along with a neglect of cereal production, have precipitated a food security crisis. To counter this crisis, each of the three countries has begun to take steps to boost its cereal production. Unfortunately, the policies adopted are further promoting cultivation of drought-prone rangelands and reduction of fallow. In Morocco, high crop prices have particularly encouraged these processes. In Algeria, these processes are explicit goals of official policy. Within Northwest Africa as a whole, a precariously high level of cereal hectarage is now perennially vulnerable to drought. In Morocco, for example, approximately 55% of the cereal hectarage is now located in 'low-rainfall' areas.

All three Northwest African countries are striving for self-sufficiency in cereal production. Drought, unfortunately, will help to make this goal an elusive dream. During the 1980s, Algeria and Tunisia each experienced five or six years of drought (depending on the definition of drought). Morocco, which was considered more fortunate, still had 'only' three or four drought years. In addition, Morocco has experienced severe droughts in 1991 and 1992. This frequency of drought suggests a significant increase in the drought hazard in the region.

Northwest Africa's increased drought hazard is primarily the result of changes in society. The region's climate has remained essentially the same during the past century. However, population levels and land-use patterns have radically changed. Farmers have progressively intensified their land use – both horizontally and vertically. This has gradually eroded traditional buffers protecting society from drought.

Meteorological drought is ultimately a relative concept, representing a shortage of rain for sustaining established human activities. As demographic growth continues to expand the quantity of grain needed to feed national populations in the region, and as population pressure and government policies result in further cultivation of marginal lands and reduction of fallow, Northwest African societies will become ever more vulnerable to drought.

The Virgin Lands Scheme in the former Soviet Union

IGOR ZONN, MICHAEL H. GLANTZ, AND
ALVIN RUBINSTEIN

To the expanding list of examples of the encroachment of human activities on marginal lands must be added the initial attempts at agricultural development of the virgin lands in the 1950s and 1960s in northern Kazakhstan and western Siberia (Figure 32). The 1964 edition of the Soviet Geographical Encyclopedia referred to such lands as 'untilled, but suitable for tillage and sowing of agricultural crops, which either never were tilled [i.e., virgin lands] or had not been cultivated for a long period of time [i.e., idle lands].' In this region, between 1954 and 1958, more than 40 million hectares of new land (larger than three times the size of Great Britain) were put into cultivation.[1] These lands straddle the northern edge of an arid zone where soils are marginal for rain-fed agricultural production. Experience later showed that such lands could not be developed for sustained agricultural production without resorting to appropriate scientifically based land-use practices.

The catalyst for the development of virgin and idle lands was the pressing need to sharply increase grain and meat production in the Soviet Union. The agricultural sector had clearly been neglected since the Bolshevik Revolution in 1917; most resources went toward industrialization efforts. With regard to increasing agricultural production, options available to the Soviet government were few: increase productivity (e.g., crop yields) on existing farmlands, such as in the highly productive Ukraine, or extend agricultural activities into seemingly unused, potential arable lands.

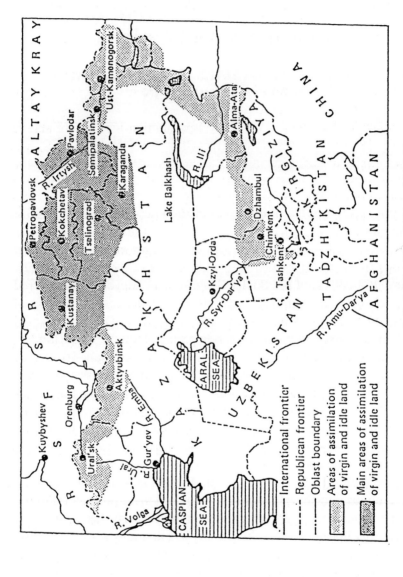

Figure 32. Virgin and idle lands of Kazakhstan, 1954–7.

The Virgin Lands Scheme was Nikita Khrushchev's pet project. Its roots can be found in the 'court' politics of the final years of the Stalin era, and in the Soviet leadership's search for a way of expanding domestic food production. Although made a member of the Politburo of the Central Committee of the CPSU (Communist Party of the Soviet Union) in 1939, Khrushchev was kept in the Ukraine until December 1949, when Stalin brought him to Moscow for political reasons.

In 1949, the Soviet Union's food production was barely able to meet the country's essential needs. Grain production, in fact, was lower than it had been in 1928, on the eve of collectivization. Soviet farmers had paid a heavy price for (1) the forced and rapid collectivization in the late 1920s and early 1930s, and (2) the dislocation of millions during the Soviet purges of the late 1930s, and the high number of casualties of World War II. As a consequence, agriculture, when compared with industry, was a relative wasteland.

From 1950 to 1964, Khrushchev chose to engage in agropolitics and played a key role in trying different policies in an attempt to raise the level of Soviet food production. His successes were evident, but exaggerated, during his period in power; only after 1964 when he was ousted by Leonid Brezhnev, whom he had promoted, were the benefits and costs of his agricultural policies to become clear and evaluated. Many of the virgin lands problems encountered during Khrushchev's tenure in office were addressed after his ouster. McCauley noted that 'the considerable successes scored by the virgin lands since 1966 underlines the wisdom [to stay in these lands].'[2]

The main climatic characteristics of the virgin lands are their continentality (i.e., hot summers, cold winters, and aridity). The degree of continentality is greater than for other parts of the Russian plain at similar latitudes. Winters are lengthy and cold with January temperatures averaging about −19 °C. Summers are short and hot with regional July temperatures averaging about 22 °C. The average length of the growing season is about 165 days. Average annual precipitation ranges from about 250 mm in the southern part of the region to 300 mm in the north; in summer evaporation is much higher than precipitation (Figure 33). Precipitation in such regions is highly variable from one year to the next, bringing a high risk to rain-fed agricultural activities.

Thus, those seeking to engage in dryland farming would be at high risk to meteorological drought with the added risk of strong dry winds during winter and spring and hot dry winds in summer. Winter winds scour snow cover off the cultivated fields, removing a protective blanket for seed and soil during harsh winters. The winds also accelerate the desiccation of soils in springtime and in summer. Thus, soil erosion is a major regional environmental problem. Earlier this century, it was generally believed that the dry winds in Kazakhstan came from Central Asia to the south. More recently, however, the view is that dry air masses originate from the Arctic north. During spring and summer when air masses shift toward the south, the warmed air masses can hold more moisture. Higher air temperatures increase evaporation, thereby reducing soil moisture. Drought-like conditions often follow.

The interaction between natural conditions in the region and human activities that do not match the ecological conditions sparks dust storms. The total number of dust storms within the steppelands of Kazakhstan ranges from 20 to 80 per year, many of which usually occur between May and July. Sdobnikov noted in 1958 that 'wind erosion of soils originated in connection with the plowing of large areas of virgin lands and the use of fine textured soils in widespread crop production. Already in regions of northern, northeastern and western Kazakhstan, especially in the Pavlodar and Aktyubinsk Regions, dust storms have damaged or completely destroyed sowings across large areas.'[3]

Precipitation varies widely from year to year throughout the region. For example, eight of the years between 1948 and 1958 were dry ones. In 1911, a dreadful drought year, precipitation was only 52 mm. In 1957, in the midst of the development of the Virgin Lands Scheme, only 9 mm of rainfall were recorded for the entire growing season. Soil temperatures can also be exceedingly high, reaching 63 °C, adversely affecting agricultural productivity.

Although physical features, such as the high degree of climate variability of the region encompassing the virgin and idle lands,

The Soviet Union Virgin Lands Scheme

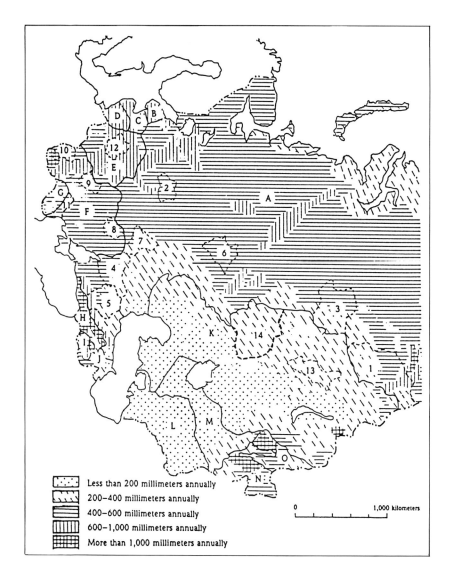

Figure 33. Rainfall map of the Soviet grain belt (from International Food Policy Research Institute Report 54, September 1986). A, Russian Federation; B, Estonia; C, Latvia; D, Lithuania; E, Belarus; F, Ukraine; G, Moldavia; H, Georgia; I, Armenia; J, Azerbaydzhan; K, Kazakhstan; L, Turkmenistan; M, Uzbekistan; N, Tadzhikistan; O, Kirghizstan. I, Altay Kray oblast; 2–14, other oblasts. Please compare with Figure 32.

may have been well understood by technical people trained in a variety of physical science disciplines, they were less understood by specialists in other fields and even less so by policymakers responsible for agricultural development. This condition exists in many parts of the world and was not unique to the former Soviet Union. Hutchinson and colleagues in an article about land use in Western Africa noted that 'although the inherent variability of climates is understood by climatologists, it seems to be routinely ignored or underestimated by developers.'[4] With regard to the Virgin Lands Scheme, Khrushchev pursued land-use policies based on his experiences in the better-watered Ukraine and on wishful thinking.

Political decisions

References to considerations of plowing the steppelands in the eastern part of the Soviet Union appeared in one of Lenin's Plans in which it was suggested that 'if farming is raised at least to the level of those parts of European Russia where climate and soils are similar, it would be possible to provide 40 to 60 million people with food.'

The first actions to develop the new lands were taken at the end of the 1920s, and particularly in 1928, when the Communist Party and the Soviet government made a historical decision to produce grain by creating state farms on previously uncultivated lands (the 'virgin lands') in several parts of the country, including northern Kazakhstan and western Siberia. Ultimately, 19 state farms in only two of Kazakhstan's northern districts – Kustanai and Petropavolvsk – were organized, covering an area of about 500 000 hectares.

The second stage of the development of the virgin lands occurred in the 1930s, when the XVIII Communist Party Congress decided to develop the country's industrial and agricultural base in the eastern regions. At that time the objective was to increase within two to three years the area of tillage by 4.5 million hectares, including more than 1 million hectares in Kazakhstan. However, the outbreak of World War II thwarted such a plan.[5]

A plan to exploit the virgin land area was proposed again in 1930. Although the plan was not implemented, a brief review of it provides some interesting history to the Virgin Lands Scheme that was ultimately pursued. Ya. A. Yakovlev, the People's Commissar for Agriculture, at the XVI Party Congress of the Bolshevik Party in the summer of 1930, proposed a program to develop 20 to 25 million hectares in Kazakhstan, western Siberia and other regions in order to cultivate wheat.[6] He suggested supplying local people with enough equipment to avoid the need to bring in large numbers of farm workers from other parts of the country: 'each human force should be used at least fifteen times more efficiently than is now done.' The model for such a plan for the virgin lands was based on Canadian farming in arid and semiarid areas. Yakovlev wrote that, as in Canada, 'the entire area should be subdivided into farms by roads passing from north to south and from east to west with each farmer tending to 200 ha of land.' If the dry soils of the Canadian prairies could successfully produce grain, why not the virgin lands with analogous soil, temperature, and rainfall conditions? The XVI Party Congress report realistically noted that 'for the time being there are no guarantees against crop failures. Guarantees should not be against crop failures but against hunger.' Thus, planners in the 1930s realized that adverse, natural environmental conditions could hamper attempts to increase food production in the region each and every year, but that those risks were worth while in an attempt to alleviate hunger throughout the country. It is important to note that the Russian fear of famine still persisted from the early 1920s when famine conditions prevailed in the Soviet Union, requiring that the new Bolshevik government seek international food assistance from capitalist countries.

In October 1948 Stalin developed his 'State Plan for Remaking Nature.' It represented a wide-ranging program focused on the exploitation of nature by, for example, constructing large hydraulic works. The part of the plan dealing with the improvement of soil quality focused only on the existing farming areas in the European territory of the Soviet Union, especially the Ukraine. Stalin, however, oblivious to the problems associated with such improvements, was intent on raising food production and of doing so within an ideological framework that was congenial to the collectivization of agriculture and the totalitarian rule that he had used to fashion Soviet society. Khrushchev had a sense of what Stalin's vision was for mobilizing and organizing the countryside in much the same way he had succeeded in doing for industry – the politics of gigantism. Khrushchev also understood that no one in the Politburo wanted to tackle such a high-risk assignment. Post-war economic difficulties, struggles between factions within the biological sciences, inadequate technical support and, not least, the death of Stalin in 1953, resulted in the shelving of Stalin's Plan.

A persistent, grave shortage of grain and meat in the Soviet Union continued into the early 1950s. This suggested that, after several decades of great technological achievements, Soviet communism could not adequately feed the Soviet people. For example, in 1953 only 31 million metric tons of grain were purchased by the state but more than 32 million tons were needed that year.[7] In this instance the government was forced to draw on its state grain reserves to cover the deficit. Sparked by such chronic food shortages, the government sought to increase crop yields in the traditional grain-producing regions in the European territory of the Soviet Union (i.e., the Soviet breadbasket) and at the same time sought to expand the area under cultivation.

When Khrushchev gained responsibility for Soviet agriculture in 1950, he proposed a program to amalgamate collective farms and villages into *agrogorods* (agricultural towns). His aims were commendable: to reduce bureaucracy, eliminate waste, encourage specialization, optimize the use of farm machinery and, in the process, enhance the Party's control over the countryside. Khrushchev's model towns, however, existed only on paper. In reality, their creation required investment, resources, personnel, and time – none of which Khrushchev had.[8] While flattering to 'Stalin's obsession with regimentation,' Khrushchev's plan soon proved to be unworkable but, with rare political perceptiveness, 'Khrushchev knew ... that any publicity is better in the long run than no publicity. He had established himself as the only man in the country besides Stalin who could act, apparently, on his own initiative and get away with it. He had *also sown the seed of an idea, which was to recur again and again after his enlargement and flower briefly in the spectacular opening of the Virgin Lands in 1954'* [italics added].[9] His main political rival, Malenkov, favored improving existing agricultural lands, whereas Khrushchev favored extending agriculture into the virgin and idle lands.[10]

At the February 1954 Plenary Session of the Central Committee of the CPSU, a report by Khrushchev was adopted 'on the further increase of grain output in the country and on the development of the virgin and idle lands.' As a direct result, a task was set forth to extend wheat cultivation in 1954–5 into 13 million hectares of the virgin and idle lands. 'The adoption of this policy in March 1954' was, according to G.A.E. Smith, 'a radical turning point in post-war Soviet agricultural policy and signified the leadership's acceptance that further economic growth could no longer be based on extracting a "surplus" from agriculture regardless of the effects on its development.'[11] A slogan used for launching this major agricultural campaign stated that 'We cannot wait for favors from nature; our goal is to take them from it!'

At the time (1954) the amount of virgin and idle lands in the Soviet Union that had been assimilated into the production system of existing cultivated land was about 19 million hectares. In an August 1954 resolution of the Central Committee of the CPSU and the Council of Ministers it was decided to bring the area sown to cereals up to 28 to 30 million hectares by 1956. The total area of newly cultivated lands was increased between 1954 and 1962 to about 42 million hectares. Smith noted that 'by 1960 the possibilities of further significant increases in sown land had come to an end (the limiting factor being the lack of precipitation in the virgin lands regions) and further increases could only be achieved by assimilating submarginal virgin land or by draining and clearing idle land in established agricultural regions.'[12]

The first detachment of farmers and ordinary citizens was sent to the virgin lands in Kazakhstan in 1954. During the first two years of the program about 650 000 new settlers went to these lands. According to Wheeler, the cultivation of the virgin lands was 'unpopular among the Kazakhs, partly because it seemed, perhaps wrongly, to threaten their traditional industry of stock breeding, and partly because it resulted in the introduction of a further 600 000 or more non-Asian immigrants.'[13]

Up to that time history had not witnessed on such a large scale the plowing up of new lands over such a short period of time. Some researchers have argued that many of the problems encountered in carrying out the Virgin Lands Scheme resulted not from what Khrushchev was trying to achieve but the rapid pace at which he tried to achieve it. Steppe regions require considerable care in their usage. The development of appropriate land tillage technologies and techniques for the purpose of rain-fed cultivation in such regions had evolved as a result of trial and error over many centuries. Yet, with the rush to cultivate virgin and idle lands, little regard was given to these traditional land-use practices or to the fragility of the soils and the harshness of the region's climate. Instead, blind faith was placed in existing technology from other agricultural regions as well as in the belief that humans could dominate nature.

In his excellent review of the Virgin Lands Scheme, McCauley succinctly stated this problem:

> It is not surprising to discover that traditional European methods were adopted when the cultivation of the new expanses in the East began. After all, the operatives came from areas with a higher level of precipitation, which results in heavier soils, so they treated the whole operation as an extension of their own traditional areas.[14]

These attitudes led to several adverse impacts such as the breakdown of the physical properties of the soil and the loss of soil fertility. Even the seeds brought into these newly cultivated areas for sowing were poorly adapted to the region's harsh environmental conditions (severe winds, heat, moisture stress).

During the first two years of the scheme (1954–6), agricultural development in northern Kazakhstan and western Siberia had both positive and negative impacts. With regard to the latter, soil degradation processes were initiated (e.g., clearing a large expanse of a sandy plain of its natural vegetation) that set the stage for the widespread, highly visible soil erosion drama that eventually took place in the virgin lands. Yet, this should have come as no surprise to Soviet leaders as history is rent with examples of failed attempts by governments to plant what they wanted to plant, where and how they wanted to plant it, with little consideration given to prevailing environmental constraints. For example, during the drought-plagued 1930s, the Canadian prairie provinces (especially Saskatchewan, which Soviet planners used as their model), witnessed Dust-Bowl-like conditions similar to those which occurred in the US Great Plains to the south. The environment of the virgin lands (and local inhabitants) would pay for the lack of attention by decisionmakers to existing environmental constraints on long-term sustainable development prospects of the virgin lands. In just a few years, following the introduction of the moldboard plow (to turn the soil and bury natural grasses of the Kazakh steppe), soils became subject to desiccation and severe wind erosion. On some days, farmers in the area could not see the sun even at midday, because of the amount of dust put into the atmosphere as a result of plowing fragile soils. The use of modern tractors only served to accelerate and intensify soil degradation on time and space scales that had not been witnessed throughout the history of farming. The plan was not realized, in large measure because of considerable political opposition to diverting capital from industrialization to the agricultural sector.

Before the large-scale development of virgin lands, erosion of the topsoil by wind action (deflation) in northern Kazakhstan was limited; it appeared as isolated minor dust storms on plowed land. In the Pavlodar Region, for example, dust storms were observed in 1922 on only one collective farm, in 1929 on about 30 000 hectares and by 1930 on even larger areas. The low level of wind erosion in the early decades of the century was because there was no widespread plowing of the land. Where it had been done, it had been as part of a fallow system which left land uncultivated for 5 to 7 years in order to restore nutrients and structure to the soils. While such a fallow system may have been less productive in economic terms, it was environmentally prudent because the risk of deflation was considerably lowered. Khrushchev paid little attention to the need for and value of letting such lands lie fallow. To him this land was not being used to its full potential. But the fragility of the soils required lengthy fallow periods if sustainable productivity was to be achieved.

Impacts on the virgin lands

As a result of widespread plowing with little regard for the fragility of the virgin and idle lands, especially on those areas with finely textured soils, the frequency and intensity of dust storms increased. Many of the storms were so large in spatial terms that they encompassed entire administrative regions in northern Kazakhstan. Dust storms robbed arable land not only of good topsoil but also of nutrients; for example, the not-infrequent loss of 3 cm of topsoil meant that the winds would have also carried away about 800 kg of nitrogen per hectare, about 200 kg of phosphorus and 6 metric tons of potassium. In the absence of human intervention it would take several centuries to replenish these soil nutrients through natural processes.

Dust storms appeared even during the first few years of plowing. As early as 1955 one could see the proverbial writing on the wall, when wind erosion caused the degradation of about 500 000 hectares of sown land with about 50 000 hectares being completely lost to agricultural production in the Kustanai Region. Deflation in 1957 on areas subjected to large-scale sowing (e.g., on state farms) reduced state farm yields to 370 kg per hectare. The average crop yield on collective farms that year was only 200 kg per hectare. In fact, crop yields throughout the virgin and idle lands were much lower than expected, which made grain produced in the region more expensive than the same amount of grain produced in other parts of the Soviet Union.

Because of the problems caused by plowing up large expanses of steppelands, local authorities had to make frequent requests to the State Planning Committee of Kazakhstan to convert the newly cultivated areas to other types of agricultural production. In fact, not an insignificant portion of these lands had become so degraded in such a short period of time that they had to be removed from grain production altogether. Nevertheless, authorities seeking to produce the best statistics (if not the best output) for their agricultural activities (e.g., with regard to area plowed, area sown, and desired yield levels) insisted on the continued plowing of virgin and idle lands, including areas with finely textured, fragile soils, and an opening up of additional lands to make up for the loss of production from the retired areas. In most cases plowing continued on areas already subjected to blowing sands as well as on lands unsuitable for farming and known to be at high risk of severe wind erosion. A constant target for continual cropping was the fallow lands. Scientists debated the various ways these lands might be incorporated on a full-time basis into crop production.

Commenting on the development of the virgin lands, Leonid Brezhnev (who at that time had been second secretary of the Central Committee of the Communist Party of Kazakhstan) stated that 'we knew, of course, that heat and aridity in that region were not unusual. But we did not yet know the ominous inexorability of the steppe calendar in which there is an especially severe disastrous drought once every ten years.'[15]

Cereal production in the Soviet Union did increase as a result of the Virgin Lands Scheme. The years 1954–8 were good years with favorable weather conditions for grain production. In fact, 1958 was a record-setting year. In the several years that followed, however, years of average crop production were interspersed with major drought-related failures.[16]

In sum, for a short time production of cereals on these newly cultivated lands increased but success was short-lived. With a variety of weather-related problems, such as unusually wet springs, early frost, increased freezing of seeds in the ground and intensification of wind erosion, parts of the newly cultivated virgin lands had to be returned for use only as livestock pastures. Plowing up grassland that had traditionally been used as rangelands had depleted nutrients from the soil, making them less productive in the future even for use by livestock. Cultivation of the virgin lands in the 1950s and early 1960s led to widespread land degradation and desertification throughout the region.

At the time (in the 1950s) many officials blamed meteorological factors for the degradation of the land and for the false start of the cultivation of the virgin and idle lands. However, human activities (e.g., plowing) had degraded the fragile vegetative cover and the surface layer of soils. At that time Izmailsky identified the cause of the drying out of the steppelands as the disappearance of the rich grass cover resulting from plowing. Plowing such soils reduces their ability to accumulate much-needed moisture. According to Russian soil scientist Dokuchaev, as quoted by P.N. Pilatov, 'there is no ground to blame climate change in the steppe region in order to explain (1) the deficiency of groundwater in the steppe area or (2) the frequent occurrence of crop failures related to drought. Only the changes in the surface properties of the steppe soils due to plowing and compaction [from heavy machinery] ... could radically change the soil moisture relationship.'[17]

Droughts in the steppes are a part of the regional climate and are to be expected. In the early years of the virgin lands project, few planners or policymakers gave any consideration to the fact that alterations in the land's surface would increase albedo (reflectivity of the Earth's surface) which in turn would alter the regional climate. Perhaps these initial large-scale, land-surface changes increased the frequency or intensity of meteorological drought in the region. Actually, some scientists did take into account the potential impacts of drought on grain production, but key people (or agricultural planners) seemed to have discounted their importance. Brezhnev wrote that 'when economic calculations were made on virgin land development, scientists considered that even if two extremely dry years were to take place during each five-year plan, we would still be able to get [favorable] grain yields in the steppe.'[18]

In the first century AD, historian and farmer Kolumella commented on persistent regional crop failures. He wrote that

> I often hear the leaders of our state accuse either the earth of barrenness or the climate for the poor and irregular yields. Some of them refer to the process as if it were a law; in their opinion the earth, tired and depleted by the luxuriant yields in ancient times, is no longer capable of providing people with the earth's previous generosity. I am sure that these reasons are far from the truth. The problem is not with heavenly wrath but with our own faults.

The euphoria generated by expected large grain outputs from the virgin and idle lands diminished quickly after 1958, in the face of recurrent drought, dust storms, and declining crop yields. It took about ten years before those planners and scientists who knew that new methods of cultivation attuned to regional environmental conditions were needed won the day. At first the Soviets looked for examples to the United States and Canada, both of which had emerged from the Dust Bowl days in the 1930s with greatly improved agricultural production systems. Zero or minimum tillage resulted in higher grain yields and reduced wind erosion and dust storms, especially during very dry years. In addition to these practices borrowed from North American agriculture, the Soviets supplemented their land-use activities with a system of tillage designed especially for the virgin lands, referred to as the conservation cropping system. Principal elements of this system were as follows: stubble mulching (preserving stubble and straw on the soil's surface), band tillage (with a distribution of bare fallow and sowing annual crops in rows perpendicular to the prevailing winds), crop rotation (grain–fallow–arable crop rotation), alternating sections of annual crops with sections of perennial grasses, sowing alternating strips of high-stem plants, and the reseeding of grassy areas that had been eroded. In addition, forest belts (called shelterbelts) would be planted (as in North America in the Dust Bowl days) to protect the stubble, band fallow and reseeded areas.

Soil-protecting cropping systems were introduced over about 20 million hectares. For example, the Tselinograd Region produced grain yields of about 600 kg per hectare before soil conservation measures had been put into place (this was the average yield from 1961 to 1965). As the system was being put into place (1966–70), yields increased to 700 kg per hectare and by 1971–3 had increased to 1190 kg per hectare.

The development of the virgin lands was of major social, economic as well as political importance. In the sparsely populated steppes of Kazakhstan, highly mechanized socialist state farms were established, settlements constructed, transportation infrastructure (rail and road ways) were built, and power plants erected. These were major accomplishments, but were achieved at great sacrifice by 'immigrants' into the region as well as by indigenous inhabitants of the steppes whose livestock herds had been displaced from their traditional rangelands. The truth is that these lands were neither virgin nor idle. According to Martha Olcott 'the term "virgin lands" was itself a misnomer, probably deliberate. The six Kazakh oblasts included in the virgin lands territory may have produced little grain but were not unexploited, for they were Kazakh pasturelands.'[19]

What happened in the virgin and idle lands of the Soviet Union in the 1950s and 1960s was in many ways a repeat of the 'Dirty Thirties' in the US Great Plains and the Canadian prairie provinces. The ensuing degradation represents a response by natural environmental processes to inappropriate land-use practices (including inappropriate technology) in agriculturally marginal areas. It represents a good example of drought following the plow in the early decades of exploitation of these 'virgin and idle' lands. Only with improved land-use practices and technology appropriate to regional environmental conditions might sustained agriculture take place. The harsh lessons of the early years of carrying out the virgin and idle lands scheme stand out as a reminder to Russian leaders and, now, the leaders of newly independent Kazakhstan that marginal lands must be treated with care and require considerable understanding before exploitation. To do less would be to set up such lands for highly destructive desertification processes.

South Africa

COLEEN VOGEL

Drought is a regular phenomenon in South Africa. Impacts of the recent drought of 1991–2 have included severe agricultural losses to commercial and subsistence farmers, reductions in reservoir levels and have exacerbated the plight of rural communities. Water provision in rural areas has been thwarted by declining water levels, the failure of boreholes, and the lack of maintenance of machinery and infrastructure. Inadequate supplies of food were also recorded.[1] It has been argued that the severity of drought impacts has been more a consequence of the mishandling of drought situations, farm management, and agricultural systems in the country than a consequence of a reduction in rainfall.

Agricultural management and physical factors have combined in fashioning the magnitude of the consequences associated with climatic extremes in South Africa. Such factors, it is argued, have also increased the degree of land abuse. 'Cultivators, often with the encouragement of their own governments, move onto the rangelands to put additional land under the plow.'[2] Several countries share such experiences, and local commentators on the recent South African drought have noted that there is 'an urgent need for climatically sensitive production management practices, drought resistant farm technologies and "nature friendly" farming systems.'[3] Thus, an understanding of the physical and societal constraints that shape both the form and function of the landscape is essential in an investigation of the relationship between farming the margins and droughts in South Africa.

South African agriculture is highly diversified and consists of two components: a predominantly commercial agricultural sector that represents mainly white agriculture (with about 20% of white farmers producing almost 80% of the total agricultural output) and a developing agricultural sector in the African bantustans (Figure 34) where farming is mainly based on subsistence production.[4] Agricultural projects are, however, contributing to an increase in the commercialization of this agricultural sector.[5]

In the following sections, the physical and socioeconomic backgrounds pertaining to South African agriculture illustrate that the migration of both commercial and rural farmers onto marginal lands has been prompted by governmental, political, and economic factors. Inappropriate farming practices have, in several instances, reduced the quality of the environment in marginal areas, thereby making communities in these areas increasingly vulnerable during droughts.

Physical factors influencing local agriculture

Both the temporal and spatial aspects of rainfall and temperature in South Africa set the limits to agricultural pursuits. The peculiar mountainous nature of the country (much of the country lies more than 1000 meters above sea level), together with the thermal influence of the oceans washing its coastline, strongly influence the temperature and moisture characteristics of the region.[6] Three rainfall regions can generally be identified: a winter rainfall area in the southwest, a coastal year-round rainfall region in the southern reaches, and a summer rainfall region.[7] Rainfall increases from the west to the east, from 125 mm to more than 1000 mm east of the Drakensberg Mountains. The 500-mm isohyet divides the country agriculturally, with the eastern areas being more suitable to regular crop production than the western regions. Because of the irregular rainfall, several areas, particularly those to the west of the country, are frequently declared to be in need of drought assistance (Figure 35). In South Africa, the influence of a variable climate on agricultural activities cannot, therefore, be underestimated.[8]

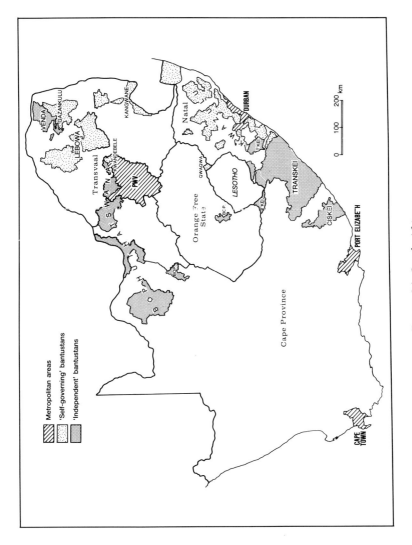

Figure 34. South Africa.

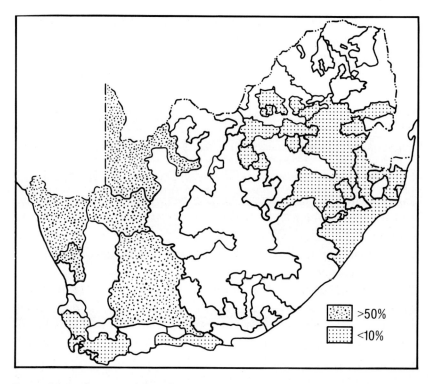

Figure 35. Declaration of drought in South Africa (1956–86) by magisterial district. Areas classified as drought-stricken in more than 50% and in less than 10% of all years are highlighted.

Features of the landscape also vary markedly (e.g., 81% of the soils are weakly weathered and calcareous) with the type of management of such soils directly affecting crop yields. The excessive use of fertilizers and water on irrigated soils has, moreover, also aggravated the erosivity of the soils and increased the problem of soil acidification. Given that only a relatively small proportion of soils can be classified as highly fertile, a conservative approach toward achieving the production potential of the agricultural sector was necessary. Despite the recognition of such need, recent assessments have shown that at least 9 million hectares of arable land and 21 million hectares of grazing lands in the white farming areas have been affected by some degree of wind and water erosion.[9]

Regional climates, soils, and topographic features combine to produce as many as 70 different types of natural vegetation (veld) in South Africa. Natural vegetation, which occupies more than 75% of the agricultural land, provides valuable fodder for much of the livestock population, as well as fuelwood, building materials, and medicinal products. Current estimates suggest that about two-thirds of the natural vegetation is in poor condition. This situation worsens when each successive drought combines with poor veld management.[10]

Agricultural activities in South Africa, including dryland maize and wheat farming and livestock raising, are largely constrained by these physical characteristics. Broadly speaking, there are seven commercial agricultural zones in the country producing a variety of products (Figure 36): the Southern and Western Cape (winter wheat, deciduous fruit, grapes); the Karroo (wool); the Eastern Cape (wheat, maize, cotton, tobacco); Natal (sugar, fruit, timber); the Orange Free State (cattle, sheep); the North and Western Transvaal (cattle, crop production), and the Highveld Region (the heartland of the agricultural economy producing grain crops, particularly maize and wheat).[11] (NB: Cape Town, Natal, the Orange Free State, and the Transvaal are politically defined provinces in South Africa. The Karroo and Highveld are geographically defined regions which transcend political boundaries.) Within these regions, there is a degree of spatial variation with regard to crop cultivation. For example, when proceeding from west to east across the country, maize is grown in three major areas: the dry west with low yields and a high risk of crop failure; the central and eastern highveld with moderate-to-high yield potential and moderate-to-low risk of crop failure; and the moist areas of Natal with high long-term yields with high input costs (e.g., for fertilizer).

There are also underdeveloped tracts of land in the country. Generations of mismanagement, ecological factors, and remoteness from metropolitan centers combine to limit intensive capital investment in agriculture in regions such as the northwestern and southeastern Transvaal.[12] In these regions several white farms are owned by absentee landlords and are unoccupied, and others have been abandoned, a phenomenon that tends to increase after severe

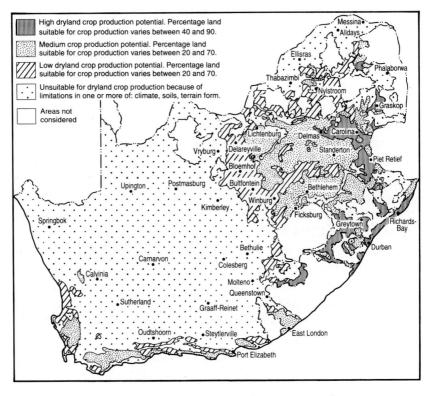

Figure 36. Overview of crop production potential.

drought; 'should avoidance of crop failure be the goal, then the western areas are least favourable [for agriculture].'[13]

The scarcity of local lands with a high potential for agricultural production has long been recognized. Although 13% of the total land area is considered to be arable, only 3% is considered to be suitable for high-potential dryland farming.[14] A considerable amount of agricultural activity is, thus, being conducted on marginal land that carries with it a high risk of failure for crop production. One estimate suggests that only one-third of white farms in South Africa are commercially viable, while the remainder are categorized as marginal.[15] These are some of the physical factors that determine the level of agricultural production. Now let us consider briefly the agricultural potential and physical constraints of black, rural agriculture.

The productive potential of the physical resources of the bantustans has been a subject of much debate. For example, some observers argue that the bantustans are as well as or better endowed with arable land than many other parts of South Africa.[16] Although it is difficult to comment conclusively on the relative percentages of high-potential farmland between the bantustans and other areas in South Africa, [17] it has been acknowledged that the agricultural potential of these areas has been influenced, often negatively, by state policies, poor rural infrastructure, and community constraints that have reduced the land's ability to support its growing populations.[18]

Thus, the bantustans have been influenced by a dual set of factors: macro-agricultural changes described earlier for commercial agriculture, and factors related specifically to the political creation of these areas. Agricultural constraints in these areas often include complex land tenure systems, inadequate farming inputs, infrastructural problems, limited access to credit, and the lack of effective lobbying in the political market.[19] Although the details of these changes cannot be elaborated here, some of the more salient features illustrate that overcultivation and expansion into the marginal lands in the bantustans have exacerbated the impacts of recurrent drought.

Socioeconomic factors shaping commercial agriculture

South African agriculture has undergone several phases of restructuring that have distinct influences on the type and quality of agricultural land that has been farmed.[20] Because of high production costs, unfavorable agricultural price trends and fluctuating yields, farmers have had to diversify and extend their operations into new areas in order to decrease their overall risk profiles.[21] A brief glimpse of the phases of agricultural developments illustrates these trends.

The first phase consisted of the initial steps in the territorial segregation of white and black farmers. Legislation, including the Land Acts of 1913 and 1936, served to confirm the existing segregation of land ownership and to abolish other forms of access to land by blacks such as sharecropping.[22] Large-scale conversion of black peasant farmers into farm laborers took place after 1913.[23] The second phase of agricultural restructuring took place after the mid-1940s and brought with it to the white farmers the mechanization of commercial farming and the introduction of high-yielding genetic technology (seeds and pesticides) and increased pressure on food production in the African homelands.

Up to the 1970s, government efforts were directed to the physical 'betterment' and upgrading of homeland agriculture through physical and technical planning.[24] The mainstay of these agricultural development efforts in the late 1970s and early 1980s included expatriate management that was introduced into the bantustans to develop agriculture and, later, adaptation whereby laborers were settled as farmers under the control of central management.[25]

In South Africa, a premium has been placed during the last 40 years on the acquisition of technology that allowed farmers to adopt a more technologically advanced approach to farming. In these years, there was little disincentive for farmers to refrain from cultivating marginal agricultural land. Inappropriate government subsidies, including tax relief on tractors and machinery and price subsidies on farm inputs such as fertilizers and interest rates, led to the over-mechanization and over-fertilization and ultimate deterioration of farmland in certain instances. Recourse to government subsidies meant that many farmers were being supported with little attention given to ensuring the mid- to long-term sustainability of their agricultural activities. In addition, state subsidies, over the long term, have enabled a small number of farmers to accumulate large amounts of land and capital.[26]

Since the beginning of the 1980s, major changes have occurred in farm policy,[27] including the shift of the agricultural sector away from regulation and subsidization toward a more competitive system based on free-market principles.[28] The combined effects in the 1980s of severe prolonged droughts, high interest rates and input prices, and poor international trade have forced the South African government to restructure agriculture in a number of 'new' ways.[29] Farmers were increasingly exposed to market interest and exchange rates from the early 1980s, budget allocations in support of white farmers have declined by approximately 50% since the late 1980s, and the real producer prices of grains have declined by more than 25% since the mid-1980s.[30] With time, however, many farmers were being increasingly pressured by price cost squeezes in the agricultural sector.[31]

Therefore, a number of farmers, including those cultivating marginal lands, have been compelled to transfer their land into the hands of commercially viable farmers.[32] Emerging from this 'shakedown' in agriculture has thus been the growth of the more successful commercial farmer, an increase in the number of part-time farmers (estimated to account for 20% of the industry), and the revival of sharecropping and labor tenancy.[33]

Although local agriculture has yielded positive benefits by providing food and fiber, foreign exchange and employment opportunities, negative effects have also occurred. Imbalances resulting from monoculture, lack of managerial expertise in hazard management, and agricultural policy have intensified the problem of degradation of natural resources.[34]

Socioeconomic factors shaping subsistence agriculture

The overall patterns and changes that have occurred in the commercial agricultural sector have also had ramifications for the rural agricultural sector. Although spin-offs and interactions between white commercial and black rural agriculture occur, the two have been described as essentially a dualistic agricultural system, with capital-intensive commercial farming existing alongside small-scale, subsistence-oriented farming.[35] One of the most glaring features of such a dichotomy in agriculture is the skewed racial access to land that has been established and perpetuated through decades of government policy and, more specifically, agricultural structuring schemes.[36] It may be said that, while drought has followed the plow in the commercial agricultural sector, in the black rural areas this process can be characterized by a process of drought following political changes in government policy.

Several writers have referred to a history of agricultural decline in black rural areas that were once self-contained.[37] Numerous factors have been cited as responsible for unsatisfactory levels of food production in the bantustans including the 'poor' decisions of local farmers and their low level of management skills, the lack of access to capital, inadequate land resources and traditional customs, and complex land tenure agreements.[38] The role of tribal authorities who control much of the land in rural areas has, for example, had a considerable influence on agricultural production in these areas.[39]

An interesting paradox also prevails in these rural areas. Many of the bantustans are plagued on the one hand by both a paucity of land for the size of the population (land starvation) and land degradation, and on the other by the fact that there are potentially arable areas left uncultivated. Reasons suggested for this phenomenon include the contrast that exists between those households that lack capital and labor but have land and those that have labor and capital but have no land.[40] The phenomenon of farmers taking jobs as migrants, if the cash income that can be derived from doing so offers a better return than farming,[41] also influences the amount of land that is farmed in these areas. Thus, the under-utilization of land in these areas has been influenced by either a lack of available capital or insufficient capital in many rural households, economics, and government policies such as 'betterment' planning.[42]

The scheme known as Betterment Planning, rooted in the Native Trust and Land Act of 1936 and reaching its height after the Tomlinson Commission of 1955,[43] officially refers to the attempts by the South African government to improve on small-scale agriculture in bantustans, combat soil erosion, and conserve the environment.[44] In order to meet these goals betterment areas were to be rehabilitated and made economically viable by being divided into residential areas, arable lands, and grazing commonage.[45] Betterment Planning is one of the external factors that has greatly influenced agricultural production and has negatively affected the physical landscape in the black rural areas.

The legacy of Betterment Planning today can be characterized by areas of increased soil erosion and reduced soil fertility. By moving people into concentrated residential areas, the plan increased rather than decreased land degradation.[46] 'Above all, years of overcrowding both people and livestock onto too little land has depleted the fertility of the land.'[47] The scheme has been sharply criticized as having failed to reach its objectives and has had little impact on an improvement of agricultural production in black rural areas.[48]

The creation of the bantustans and the movement of people into already congested labor reserves aggravated existing pressures on the land. Population density and demands on rural land resources consequently soared. In Natal, for example, the population density on white farms in 1980 was 22 people per square kilometer, whereas in Kwa Zulu it was 76 people per square kilometer.[49]

Historical evidence for drought follows the plow: state drought and veld management policies 1923–92

Over the years, several national drought and agricultural commissions and investigations have repeatedly drawn attention to problems generated by the exploitation of vulnerable resources. As previously indicated, many of the farming activities conducted in South Africa have been on land with low-to-medium agricultural potential. The results of several government commissions and investigations of white commercial agriculture have indicated that sustainable agriculture was not being practiced in the country and that inappropriate farming practices were as much to blame for the loss of agricultural production during drought periods as were the physical aspects of the weather (i.e., meteorological drought).

At least as early as 1923, the issue of whether drought impacts were tied only to climatic variations was addressed by a Drought Investigation Commission:

> Whether the character has altered or not, or its quantity diminished, drought losses can be fully explained without presuming a deterioration in the rainfall. Your Commissioners had a vast amount of evidence placed before them from which only one conclusion can be drawn, namely, that the severe losses of the 1919 drought were caused principally by faulty veld and stock management.[50]

Subsequent investigations made it increasingly clear that mismanagement and over-exploitation of the land (veld) were aggravating the impacts of droughts:

> the possibility of trekking with stock to better grazing in times of severe drought is becoming increasingly limited.[51]

and

> In many respects, the recent drought laid bare the very roots of many shortcomings and defects in farming. Many of the problems and shortcomings which are mentioned in this report and are specifically related to drought, are equally related to the shortcomings and problems affecting agriculture in general.[52]

The size of farms as well as the quality of the land have also played roles in heightening the impacts of droughts. For example, poor land quality, an intensification of farming, and harsh drought impacts have progressively burdened smaller farmers. 'Overgrazing occurs when smaller farmers and farms with high capital investment endeavor to make a financial success of their operations. This practice, however, results in the rapid development of grazing problems during times of low rainfall.'[53]

The culmination of official concern expressed in agricultural investigations prior to the 1960s was reached in the late 1960s with the comprehensive and extensive Marais–Du Plessis Commission. An important conclusion of this investigation was that the smaller, uneconomical farming units be eliminated to ensure better conservation farming. In addition, the Commission noted that

> attention will have to be given in good time to measures for adapting production systems to environmental conditions and for making agriculture less vulnerable to periodic droughts.[54]

The Marais–Du Plessis Commission noted that the state subsidization policy had failed. Government subsidy policies have intensified the agricultural exploitation of marginal land, thereby increasing the vulnerability of such farmers to drought. Rather than assisting farmers, agricultural subsidies had ensured that poor farmers (i.e., those using inappropriate farming methods) were able to stay on the land.[55] In addition, to maintain security in the border areas of Zimbabwe and Botswana, millions of South African rand have been given to white farmers in the form of subsidies to discourage absenteeism. In 1980 and 1985, the government provided R20.9 million to 212 farmers and R9.2 million in the form of 123 new loans to agricultural companies in these areas.[56] Whereas farming of such land should involve low-input farm systems, government assistance programs have resulted in the generation of high levels of inputs and growing farmer debt.

Several factors had combined to increase farming in agriculturally marginal areas, including state legislation, land ownership, subdivision of farm land, abuse of the land by over-usage of chemicals and fertilizers, and state subsidies. Heavy subsidization of white commercial farming, at the expense of black farmers, prevented farmers from extricating themselves from their situation.

The case of black rural areas

Droughts, rather than causing agricultural and food problems in black rural areas of South Africa, have often brought the parlous state of agriculture into sharper focus.[57] Several examples of the consequences of farming at the margins and the exacerbation of the effects of 'normal' droughts exist for most of the bantustans including the Transkei,[58] Ciskei,[59] Lebowa,[60] Kwa Zulu,[61] and Bophuthatswana.[62] Although several of these areas cannot be classified as having been marginal in the past in terms of agricultural potential, the case of the Transkei shows that they have become increasingly marginal.

In terms of soils, vegetation and other agro-ecological factors, Transkei has a high and diversified agricultural potential.[63] The environmental conditions of the land are, however, undergoing marked deterioration, particularly in the form of increased soil erosion. Such conditions in turn adversely influence crop yields; for example, there is strong statistical evidence that maize yields have been declining.[64] The impacts of meteorological droughts have been superimposed on these trends (in 1981 approximately 50 000 metric tons of maize were harvested. The harvest was reduced by half in 1982 at the height of drought).[65] Transkei had been a net exporter of maize at the turn of the century but has recently become an importer of grain.[66]

Livestock farming is also an important household asset in the Transkei. The national herd size has remained fairly stable over the past 30 years, apart from the harsh droughts of the 1980s and

1990s.[67] Of importance to this discussion is the fact that local physical resource endowments have not been shown to be a problem in the Transkei but, rather, the reason for progressive land deterioration is found in institutional factors such as inappropriate land-use planning (e.g., Betterment Planning), and lack of support services, farming systems, finance and farming equipment.[68]

In their analysis of the impacts associated with the 1980s drought Bolus and Miller suggested that

> drought and its effects have to be located against a background of rural undevelopment and population influx controls which have resulted in a sustained rural crisis. The drought has accelerated all the symptoms related to poverty whether by lawlessness or malnutrition but has not caused them. These must be traced to the structural characteristics of racial capitalist development in South Africa which concentrates poverty along racial, spatial and sexual lines.[69]

Panaceas to the problem: success or failure?

In the spirit of the Marais–Du Plessis Commission, three programs were established to redress the problems of land deterioration, farming in marginal areas, and heightened drought consequences: (1) The National Grazing Strategy (NGS); (2) the conversion of crop land to grazing land; and (3) recent agricultural assistance measures to farmers' sector.

(1) *The National Grazing Strategy.* The denuding of the vegetative cover in grazing areas has always been a problem for livestock owners. A 1986 survey of more than 1000 farmers from several agricultural centers throughout the country clearly illustrated that 'drought follows the plow' was the order of the day, at least in the minds of the farmers. When questioned about their impressions of the main causes of land degradation and deterioration, an overwhelming 87% stated that overgrazing was the major problem, followed by drought.[70]

During the severe drought of the 1980s, the critical importance of grazing land was again evident. In a survey about cattle, water, and grazing conditions for a number of districts the heightened problem of overgrazing, as opposed to lack of water, was noted in several areas of the country (Figure 37).[71] This survey illustrated clearly that, in terms of veld management, drought does indeed follow inappropriate farming methods and that effective veld management would offset this problem.

During the past decade, official plans have been implemented to halt the degradation of South African rangelands. In May 1984, the White Paper on Agricultural Policy made reference to the alarming deterioration of rangeland resources in the country.[72] As a result of this document, the NGS emerged and a number of aims were stated, including the 'wise' use and management of natural and cultivated pastures, and the sustainable use of land by farming according to the regional and local climatic, soil, and vegetation conditions.

Important policy aspects contained in the NGS included the provision that drought disaster relief be given only to those land owners known to be practicing conservation measures. The NGS once again drew attention to the problem of uneconomic farm size and poor farm management, as well as to the fiscal policy and marketing arrangements that were obstacles to achieving the objective of the NGS.

(2) *Land Conversion Scheme.* Complementing the National Grazing Strategy is the Land Conversion Scheme (established in October 1987). The scheme was designed to convert marginal farmland in the summer rainfall grain areas to grazing land. Reasons cited for land conversion included the increased high risk of investment in farming because of the financial costs of farming in marginal areas that are increased by drought years, the increasing production costs that reached a critical threshold in the mid-1980s, and government policy changes that reduced subsidies to farmers. Farming in marginal agricultural areas has become increasingly discouraged.

The benefits of the Land Conversion Scheme include improved grazing capacity, low cost in terms of farmer overheads (as opposed to maize production), and the withdrawal of approximately 1 million hectares from maize production,

CASE STUDIES AND CONCLUSIONS

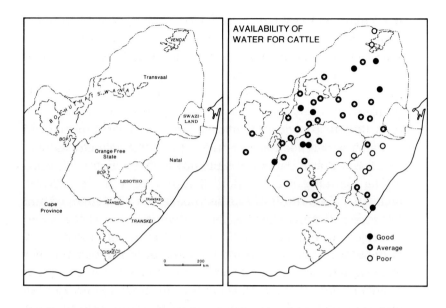

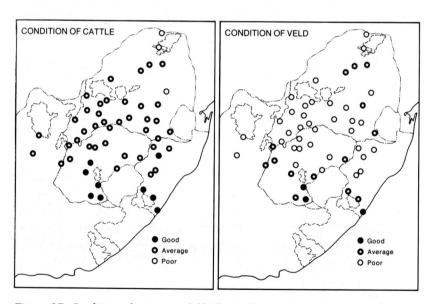

Figure 37. Condition of water available for cattle, condition of cattle, and condition of veld.

particularly in marginal areas.[73] Some of the problems with this plan include specific elements of the scheme, the expense of such a project, and the creation of short-term cash flow deficits to farmers.[74] Although the program has been voluntary, it has nevertheless gained some degree of popularity (Figure 38).

(3) *Agricultural assistance measures.* Soaring farmer debt reached a peak (approximately R20 billion) with the recent drought of 1991–2. As a result of the predicament in which many farmers found themselves, the state launched a comprehensive revised assistance scheme.[75] Some of the principles of the scheme include the optimal utilization of the scarce agricultural resource and the removal of state guarantees from the system, thereby returning to a more market-oriented agriculture. Part of the scheme also includes a refunding of unmanageable debt to those 'farmers who followed sound and correct farming and financing procedures.'[76] Although this scheme is designed to promote sustainable agriculture and has made provisions for

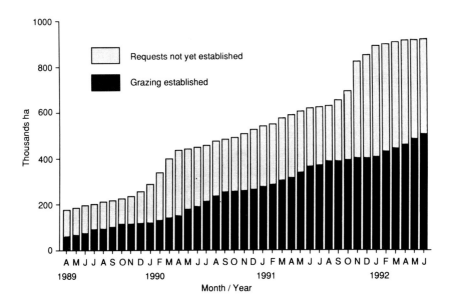

Figure 38. Numbers of requested and accepted cases for the Land Conversion Scheme.

those vast areas where rural, subsistence agriculture occurs, it is this sector of the population where the farming of marginal areas is more pronounced.

Have lessons been learned?

The recurrence of severe drought in the early 1990s has brought with it a number of deleterious impacts. The drought came at a time when local political and economic instabilities prevailed throughout the country. An estimated 50% of the national population live below the poverty line, 40% of working age have no formal jobs, and 10% of new job entrants cannot find a job.[77] As in previous cases, the impacts of drought in the early 1990s have been worse for the poor, rural communities, as opposed to commercial agriculture.[78] Engineers, called in to assist with recent drought relief efforts, have noted that the drought's impacts, particularly in the rural areas, were not the direct result of drought but rather the result of years of neglect, lack of investment, and lack of maintenance.[79]

The combined efforts of state and non-governmental organizations, collectively known as the National Consultative Forum on Drought, established in 1992, have seen the scope of drought relief widening to include all those sectors of the population that had been negatively affected by the drought. This recent drought generated a greater awareness of and appreciation for the 'roots' of the problem of drought and its adverse environmental and societal consequences in South Africa. Such heightened awareness should make the government better equipped to mitigate the impacts of future droughts throughout the country.

Conclusions

The consequences of recurrent drought episodes in South Africa have clearly been exacerbated by both political and socioeconomic factors. The farming of large stretches of land that are in fact marginal for sustained agricultural production and the extension into marginal areas of otherwise suitable agricultural practices have been largely initiated by economic factors that have encouraged farmers to seek relatively short-term economic benefits at the mid- and long-term expense of the physical environment. Environmental degradation of farmland, highlighted as well as accelerated during droughts, can thus in large measure be ascribed to inappropriate agricultural policies and land-use practices. When droughts occur, the legacy of past unsound farming practices becomes apparent, as farmers struggle with declining yields and spiraling debts. State policies exhibited in the major disparities in assistance between white and black farmers have also resulted in deleterious land-use practices.

Droughts, of varying duration and spatial extent, are recurrent phenomena in South Africa, with a drought occurring somewhere in the country each year. Droughts, moreover, expose inappropriate facets of agricultural management and socioeconomic policy, often leaving in their wake a substantial expansion of land increasingly marginal for agricultural production. They also increase farmer debt.

For the commercial agricultural sector, drought together with economic factors such as producer price variations, high interest rates, and cost–price squeezes have decreased the opportunities for successful farming in South Africa. In the past farmers were encouraged through state subsidies to exploit marginal farming areas. Recent restructuring within the agricultural sector, along with a re-evaluation of policies related to subsidy allocation, have discouraged farming these areas and, in certain cases, it is no longer a viable option.

Agriculture in the bantustans has been affected by numerous different socioeconomic and political constraints as well as a deteriorating landscape and an increased marginality of the people as well as the land. The various resettlement, betterment, and other politically designed schemes have contributed to the deterioration of the land. Drought, particularly a severe drought, merely exposes a number of vulnerabilities and weaknesses faced by those who live in the black rural areas.

Drought following the plow is not a new phenomenon in South Africa. Reference to increased degradation of the biophysical landscape has been noted at least since the early decades of the 1900s in several national drought commission reports. However, major socioeconomic and political changes in the new South Africa have initiated changes in thinking about the agricultural sector and about better management of droughts through sustainable development practices.

Several strategies have been devised in the past to halt the spread of farming into marginal lands and the degradation of those environments. These have, however, been met with opposition in certain instances, because they are not considered to be cost-effective and because such strategies often serve to placate farmers during harsh periods only to be followed by a return to old habits on the part of farmers and government when the rains return.

Notwithstanding this 'drought follows the plow' situation and inappropriate local policies, effective farm management and the adoption of sustainable agricultural practices can reverse the existing situation. Discriminate use of fertilizers, allowing the land to 'rest-over,' and alternative plowing procedures and foraging schemes can reduce the stress placed on the landscape and, hence, reduce the pressure to over-extend cultivation into marginal agricultural areas. Recent success stories of high yields and the identification of more sustainable agricultural methods in the face of drought are testimony to such strategies.[80]

Is the stork outrunning the plow?

MICHAEL H. GLANTZ

In a book on world hunger, written in 1975 in response to the famines in West Africa, Ethiopia and Somalia, W. Stanley Mooneyham addressed the population issue in a chapter entitled 'Is the Stork Outrunning the Plow?'[1] Are we producing people faster than we can devise ways to feed them? The population issue has long been a sticky problem for governments, rich and poor alike.

It is not so easy to convince people that family planning is in their (and the environment's) best and long-term interests. In large measure children are a part of family insurance against adversities in the future, especially in developing countries. The more children, the better the chance for survival of the whole family. Yes, there will be more mouths to feed – but there will also be more hands to work in the fields or with the herds. There will be a higher probability that some of the children will survive to adulthood and will take care of their aging parents.

Many people in Third World countries believe that calls for population control are a tactic of the richer nations to maintain their economic domination of global resources. Some countries, such as China, view their large and expanding populations as assets that can be used to enhance their country's development prospects.

Many countries today are unable to supply their citizens with an adequate level of nutrition. Children and the elderly are among the first to suffer. And this is the case in times of good weather with rainfall amounts adequate for agricultural production. In times of extreme meteorological events (such as droughts or floods), food production on rain-fed agricultural (and grazing) lands declines, which, in turn, prompts a decline in the nutritional status of children and others. Runs of drought years wreak considerable havoc by undermining any chance governments might have had at achieving a modicum of food security.

To make matters even worse, people around the globe, representing all kinds of cultures and technologies, are moving into areas considered in the past as too marginal for sustained agricultural production. The consequences of such movements have been masked by brief periods of good rains, by reduced fallow time, and even by naturally occurring droughts that suggest that nature has created devastation in the margins.

The reasons for the extension of human activities into the margins are varied. They can generally be divided into 'push' and 'pull' factors. Some of these factors are natural, while others are anthropogenic. The set of factors prompting marginal land utilization vary from one location and point in time to another. Population numbers alone do not provide the true cause of the degradation of marginal lands, although they are directly implicated in the process of degradation.

Almost 200 years ago Malthus presented his treatise on how population growth rates would eventually overtake our ability to feed the expanded population. Many subsequent studies have attempted to determine the Earth's carrying capacity; how many people can the resources of the Earth support? Estimates have ranged into the tens of billions by some scientists to less than ten billion by others. Little consideration is given, however, to the quality of life awaiting the majority of those tens of billions. Even with today's six billion or so inhabitants, a large majority are poorly fed, clothed, and housed.

A missing element from Malthus's treatise centers on the fact that all land has not been created equal; that is, some lands are more fertile than others. Thus, the Malthusian prospects of the inability of the human race to feed itself did not take into account the fact that of the relatively small amount of the Earth's landmass that is arable, only a small percentage is highly fertile, with promising potential for long-term sustainability. As it does not require the services of a weatherman to know which way the wind blows, an expert is not required to see that limits need to be placed on human activities that degrade the Earth's natural resource base, if a widespread collapse of food production processes in the future is to be avoided.

The finiteness of the Earth's resources was a popular issue in the late 1960s and early 1970s.[2] Those who have a faith in technology, however, believe that those natural limits to growth can be manipulated by human intervention.[3] Others blame technology and human intervention in general for the downward spiraling of environmental quality.[4]

Technology has been viewed by many as the way out of this dilemma. It can be used to slow down population growth, produce more foodstuffs, restore destroyed land surfaces, bring water to deserts and, through bioengineering, create manufacturing processes that can produce edible products. But those in need of food and those countries at risk to severe and chronic food shortages do not have the resources to develop or to buy such technologies. In other words, those in need do not have the wherewithal to pay for technologies required to satisfy those needs.

World-famous economist Kenneth Boulding wrote about the dismal theorem in economics. 'If the only thing that can check the growth of population is starvation and misery, then the population will grow until it is sufficiently miserable and starving to check its growth.' He then proposed the 'utterly dismal theorem,' which says that 'if the only thing which can check the growth of population is starvation and misery, then the ultimate result of any technological improvement is to enable a larger number of people to live in misery than before and hence to increase the total sum of human misery.'[5]

Societies must confront the underlying causes of population movements into their remaining marginal areas. It is dangerous to see these areas as the remaining frontiers for human settlements in future decades. Recent history has shown that these lands are extremely fragile. Once destroyed, they can only be restored at great expense, which no government or aid agency can afford (let alone has the will to do). It is also important to use productive lands with great care. Degradation of prime agricultural or grazing lands also puts pressure on the margins, as lost productivity has to be made up in other places. It is cheaper and easier to maintain the productivity of good lands than to try to restore them once they have been degraded.

The linkage between poor people and degraded environments (lands at risk or marginal lands) does exist. But the linkage requires close scrutiny to determine whether poor people are creating those environments or whether the poor environments are keeping the people in them poor.[6] What role do outsiders (those outside the marginal areas) play in the movement of human activities into and degradation of the margins?

In sum, one could effectively argue that the stork will eventually outrun the plow, as originally suggested by Malthus, and that droughts will continue to follow the plow. A recent article in the *Economist* on the newly independent states of Central Asia (Kazakhstan, Turkmenistan, Kyrgyzstan, Tajikistan, and Uzbekistan) focused on population growth rates and finite, if not dwindling, natural resources in the region. The author referred to the region as 'Malthustan.'[7] The variety of high, dry, and cold marginal areas around the globe could also be encompassed by the fictional country of Malthustan.

Today there is great interest in the concept of sustainable development. Sustainability requires the use of resources in such a way as to avoid depletion of the resources base. Using a banking analogy, we should conscientiously decide to live off the interest instead of the principal.

The idea that 'drought follows the plow' can be instructive in that it warns planners and policymakers (even those wearing rose-colored glasses) that there are adverse consequences to developing new agricultural lands and rangelands, if those plans are not carefully done. One cannot expect, as was shown by the Virgin Lands Scheme in the former Soviet Union in the mid-1950s, that the environment can be easily manipulated by political decisions or ideological stances. Careful planning and assessments must be undertaken well before populations are encouraged to move into 'marginal' areas. Careful consideration must be given to the long-term implications for the environment and its ability to sustain a government's active encouragement of the movement of farmers and herders into lands that may be marginal for sustained agricultural production.

We now hear more and more about the possibility of global climate change. Speculation abounds about how such a change might affect regions, countries, ecosystems, and resources around the globe. Will there be more droughts or less? Will they be more intense than those of today? Will rainfall variability increase or decrease with a climate change? Will there be climate change surprises (events that are the opposite of what one might reasonably expect)? But, as of today, it is still in the speculative phase, and such speculation must be used with great care. With such a climate change, some of these agricultural margins might in time prove to be better than marginal, while the present good, fertile lands may eventually fall into the marginal category. An improved understanding of human interactions with the environment in the marginal areas is essential to an improved understanding of how societies might best respond to the regional impacts of global climate change decades in the future.

Notes

Preface

1. Benedick, R.E., 1991: *Ozone Diplomacy: New Directions for Safeguarding the Planet*. Cambridge, Massachusetts: Harvard University Press.
2. WCED (World Commission on Environment and Development), 1987: *Our Common Future*. New York: Oxford University Press.
3. Kempton, W., 1991: Lay perspectives on global climate change. *Global Environmental Change*, **1**, 3, 183–208.
4. For policymakers, see Lewis Harris & Associates, 1989: *Public and Leadership Attitudes to the Environment in Four Continents*. New York, NY: Lewis Harris & Associates. For scientists, see Stewart, T.R., Mumpower, J.L. & Reagan-Cironcione, P., 1992: *Scientists' Agreement and Disagreement about Global Climate Change: Evidence from Surveys*. Albany, New York: Center for Policy Research, State University of New York.
5. For example, see Pearce, F., 1992: American sceptic plays down global warming fears. *New Scientist*, 26 December, 6.
6. IPCC (Intergovernmental Panel on Climate Change), 1992: *Climate Change: The 1990 and 1992 Assessments*. Geneva, Switzerland: IPCC Secretariat.
7. Bryson, R. & Murray, T.J. 1977: *Climates of Hunger*. Madison, Wisconsin: University of Wisconsin Press.
8. Glantz, M.H. (ed.), 1977: *Desertification: Environmental Degradation in and around Arid Lands*. Boulder, Colorado: Westview Press.
9. See, for example, Glantz, op. cit., or Glantz, M.H. (ed.), 1988: *Societal Responses to Regional Climatic Change: Forecasting by Analogy*. Boulder, Colorado: Westview Press.

Introduction

1. See, for example, Glantz, M.H. (ed.), 1977: *Desertification: Environmental Degradation in and around Arid Lands*. Boulder, Colorado:

Westview Press, and Glantz, M.H. (ed.), 1988: *Societal Responses to Regional Climatic Change: Forecasting by Analogy.* Boulder, Colorado: Westview Press.

Drought, desertification and food production

1. Schneider, S.H., with Meisrow, L.E., 1976: *The Genesis Strategy: Climate and Global Survival.* New York: Plenum Publishing Corporation.
2. See, for example, Wilhite, D.A. & Glantz, M.H., 1985: Understanding the drought phenomenon: the role of definitions. *Water International,* **10,** 111–20, or Glantz, M.H. & Katz, R.W., 1977: When is a drought a drought? *Nature,* **267,** 192–3.
3. Sanford, S., 1978: Towards a definition of drought. In: M.T. Hinchey (ed.), *Symposium on Drought in Botswana.* Hanover, New Jersey: University Press of New England.
4. See, for example, Garcia, R., 1981: Drought and man: the 1972 case history. Vol. 1, *Nature Pleads Not Guilty.* New York: Pergamon Press; Watts, M., 1983: *Silent Violence.* Berkeley, California: University of California Press.
5. Morse, B., 1987: Foreword. In: M.H. Glantz (ed.), *Drought and Hunger in Africa.* Cambridge: Cambridge University Press, xiii–xx.
6. ICIHI (Independent Commission on International Humanitarian Issues), 1985: *Famine: A Man-Made Disaster?* New York: Vintage Books.
7. For the US Great Plains, see Bowden, M.J., Kates R.W., Kay, P.A., Riebsame, W.E., Warrick, R.A., Johnson, D.L., Grould, H.A. & Weiner, D., 1981: The effect of climate fluctuations on human populations: two hypotheses. In: T.M.L. Wigley, M.J. Ingram & G. Farmer (eds.), *Climate and History: Studies in Past Climates and Their Impact on Man.* Cambridge: Cambridge University Press, 479–513. For the West African Sahel, see Sircoulon, J., 1976: Les données hydropluviometriques de la sécheresse récente en Afrique intertropicale: comparison avec les sécheresses 1913 et 1940. Cahiers ORSTOM, **13,** 75–174, or Kates, R.W., 1981: Drought in the Sahel: competing views as to what really happened in 1910–14 and 1968–74. *Mazingira,* **5,** 2, 72–80.
8. Hall, A., 1978: *Drought and Irrigation in North-East Brazil.* Cambridge: Cambridge University Press; Shepherd, J., 1975: *The Politics of Starvation.* New York: Carnegie Endowment for Peace; Watts, M., 1983: *Silent Violence.* Berkeley, California: University of California Press.
9. Glantz, M.H., 1990: *On Assessing Winners and Losers in the Context of Global Warming.* Report of a workshop 18–21 June 1990 in St Julians, Malta. Boulder, Colorado: National Center for Atmospheric Research.

10. Glantz, M.H. & Orlovsky, N., 1983: Desertification: a review of the concept. *Desertification Bulletin*, **9**, 15–22.
11. Mainguet, M., 1991: *Desertification: Natural Background and Human Mismanagement*. Berlin: Springer-Verlag, 165.
12. Baker, R., 1984: Protecting the environment against the poor: the historical roots of the soil erosion orthodoxy in the Third World. *The Ecologist*, **14**, 53–60.
13. Charney, W.J., Quick, S.-H. & Kornfield, J., 1977: A comparative study of the effects of albedo change on drought in semiarid regions. *Journal of Atmospheric Science*, **34**, 1366–85.
14. Otterman, J., 1977: Anthropogenic impact on the albedo of the earth. *Climatic Change*, **1**, 137–55; Chervin, R.M., 1979: Response of the NCAR atmospheric general circulation model to changed albedo. *Report of the JOC Study Conference on Climate Models: Performance, Intercomparison and Sensitivity Studies*, held 3–7 April 1978 in Washington, DC. GARP Pub. Ser. No. 22, 563–81. Geneva, Switzerland: WMO; Kellogg, W.W. & Schneider, S.H., 1977: Climate, desertification and human activities. In: M.H. Glantz (ed.), *Desertification*. Boulder, Colorado: Westview Press, 141–64; Dickinson, R.E., 1980: Effects of tropical deforestation on climate. In: *Blowing in the Wind: Deforestation and Long-Range Implications*. Studies in Third World Societies Publication No. 14. Williamsburg, Virginia: Department of Anthropology, College of William and Mary; Charney *et al.*, op. cit.; Anthes, R.A., 1984: Enhancement of convective precipitation by mesoscale variations in vegetative covering in semiarid regions. *Journal of Climate and Meteorology*, **23**, 541–54.
15. Charney *et al.*, op. cit.; Courel, M.F., Kandel, R.W. & Rasool, S.I., 1984: Surface albedo and the Sahel drought. *Nature*, **307**, 528–30.
16. Tucker, C.J., Dregne, H.E. & Newcomb, W.W., 1991: Expansion and contraction of the Sahara Desert from 1980 to 1990. *Science*, **253**, 299–301.
17. J. Shukla, University of Maryland, personal communication, 1991.
18. Balling, R.C., Jr, 1991: Impact of desertification on regional and global warming. *Bulletin of the American Meteorological Society*, **72**, 232–4.
19. In a recent article about 'carrying capacity,' David Norse noted that 'a region's human-carrying capacity can be increased by raising land and labor productivity or through trade with better endowed regions.' (Norse, D., 1992: A new strategy for feeding a crowded planet. *Environment*, **34**, 6–11, 32–9.)
20. See, for example, Brown, L.R., 1968: *Seeds of Change*, New York: Praeger Press; Mellor, J.W. & Adams, R.H., 1984: Feeding the underdeveloped world. *Chemical Engineering News*, **62**, 32–9.

21. Tolba, M.K., 1984: Soil erosion threatens world agriculture. *Mazingira*, **8**, 7.
22. Anon., 1985. Erosion, drought and deserts. *The Courier (UNESCO)*, January, 8.
23. Cross, M., 1984: The failure of Africa's agriculture. *New Scientist*, **102**, 5.
24. Norman, C., 1985: The technological challenge in Africa. *Science*, **227**, 8 February, 616–17.
25. For example, see Brown, L.R., 1981: World population growth, soil erosion, and food security. *Science*, **214**, 995–1002; or Norse, op. cit.
26. UN FAO, 1981: Agriculture: towards 2000. Rome, Italy: United Nations Food and Agriculture Organization (UN FAO), 126.
27. Biswas, M.R. & Biswas, A.K., 1979: *Food, Climate and Man*. New York: Wiley, 195.
28. US OTA (Office of Technology Assessment), 1984: Africa tomorrow: issues in technology, agriculture, and US foreign aid. Washington, DC: US Government Printing Office, 21.
29. Cross, op. cit.; see also Earthscan, 1984: *Environment and Conflict*. Earthscan Briefing Document No. 40. London, UK: International Institute for Environment and Development, 12.
30. Schulz, A., 1982: Reorganizing deserts: mechanization and marginal lands in Southwest Asia. In: B. Spooner & H.S. Mann (eds.), *Desertification and Development: Dryland Ecology in Social Perspective*. New York: Academic Press, 34.
31. Webb, W.P., 1931: *The Great Plains*. New York: Grosset & Dunlap, 340.
32. Kutzleb, C.R., 1968: *Rain Follows the Plow: The History of an Idea*. PhD Thesis. Boulder, Colorado: University of Colorado.
33. Webb, op. cit., 340.
34. Ibid., 377.
35. Ibid., 376.
36. Smith, H.N., 1950: *Virgin Land: The American West As Symbol and Myth*. New York: Vintage Books, 201.
37. Spence, C.C., 1980: *The Rainmakers: American 'Pluviculture' to World War II*. Lincoln, Nebraska: University of Nebraska Press, 7.
38. Kutzleb, op. cit., 10.
39. Worster, D., 1979: *Dust Bowl: The Southern Plains in the 1930s*. New York: Oxford University Press.
40. For example, see Great Plains Committee, 1936: *The Future of the Great Plains*. Washington, DC: US Government Printing Office.
41. See, for example, Kellogg, C.E., 1935: Soil blowing and dust storms. Miscellaneous Publication No. 221. Washington, DC: US Department of Agriculture.

42. For example, see Chapter 3 in Bittinger, M.W. & Green, E.B., 1980: *You Never Miss the Water Till...* Littleton, Colorado: Water Resources Publications.
43. Kessler, E., Alexander, D.Y. & Rarick, J.F., 1978: Duststorms from the US High Plains in later winter 1977: search for cause and implications. *Proceedings, Oklahoma Academy of Sciences,* **58,** 116–28.
44. Bernard, H.W., Jr, 1980: *The Greenhouse Effect,* Chapter 4. Cambridge, Massachusetts: Ballinger Publishers.
45. For example, see Warrick, R.A., 1984: The possible impacts on wheat production of a recurrence of the 1930s drought in the US Great Plains. *Climatic Change,* **6,** 1, 5–26; or Rosenberg, N.J. & Crosson, P.R., 1991: The MINK project: a new methodology for identifying regional influences of, and responses to, increasing atmospheric CO_2 and climate change. *Environmental Conservation,* **18,** 4, 313–22.
46. Matthews, A., 1990: The Poppers and the plains. *The New York Times Magazine,* June 24.

The West African Sahel

1. Morentz, J.W., 1980: Communications in the Sahel drought: comparing the mass media with other channels of international communication. In: *Disasters and the Mass Media,* Proceedings of the Committee on Disasters and the Mass Media Workshop, February 1979. Washington, DC: National Academy of Sciences.
2. For some examples, see Glantz, M.H. (ed.), 1976a: *The Politics of Natural Disaster: The Case of the Sahel Drought.* New York: Praeger Publishers; Garcia, R., 1981: Drought and man: the 1972 case history. Vol. 1, *Nature Pleads Not Guilty.* New York: Pergamon Press; or Watts, M., 1983: *Silent Violence.* Berkeley, California: University of California Press.
3. Temple, R.S. & Thomas, M.E.R., 1973: The Sahelian drought: a disaster for livestock populations. *World Animal Review,* **8,** 3.
4. Lamb, P.J., 1982: Persistence of Subsaharan drought. *Nature,* **299,** 36–47; Nicholson, S.E., 1981: Rainfall and atmospheric circulation during drought periods and wetter years in West Africa. *Monthly Weather Review,* **109,** 2191–208; Lamb, P.J. & Peppler, R.A., 1991: West Africa. In: M.H. Glantz, R.W. Katz & N. Nicholls (eds.), *Teleconnections Linking Worldwide Climate Anomalies.* Cambridge: Cambridge University Press, 121–90.
5. Winstanley, D., 1985: Africa in drought: a change of climate? *Weatherwise,* **38,** 74–86.

6. Glantz, M.H. & Katz, R.W., 1985: Drought as a constraint in sub-Saharan Africa. *Ambio*, **14**, 334–9.
7. Glantz, M.H., 1992: Global warming and environmental change in sub-Saharan Africa. *Global Environmental Change*, **2**, 9, 183–204.
8. Garcia, R., 1981: Drought and man: the 1972 case history. Vol. 1, *Nature Pleads Not Guilty*. New York: Pergamon Press, 183.
9. Campbell, D.J., 1977: Strategies for coping with drought in the Sahel: a study of recent population movements in the Department of Maradi, Niger. PhD Dissertation. Worcester, Massachusetts: Clark University, 79.
10. Warren, A. & Khogali, M., 1992: *Assessment of Desertification and Drought in the Sudano-Sahelian Region, 1985–1991*. New York: UN Sudano-Sahelian Office (UNSO), ix.
11. Campbell, op. cit., 173.
12. Ibid., 91.
13. de Wilde, J.C., 1984: *Agriculture, Marketing, and Pricing in Sub-Saharan Africa*. Los Angeles, California: UCLA African Studies Center, 4.
14. Franke, R.W. & Chasin, B.H., 1980: *Seeds of Famine*. Montclair, New Jersey: Allanheld, Osmun & Co.
15. Campbell, op. cit., 96.
16. Lofchie, M.F., 1990: Kenya: still an economic miracle? *Current History*, **89**, 209–22.
17. Shepherd, J., 1975: *The Politics of Starvation*. New York: Carnegie Endowment for Peace.
18. Campbell, op. cit., 80.
19. Ibid., 81 (C. Reynaut as quoted in Campbell).
20. Ho, T.J., 1990: Population growth and agricultural productivity. In: G.T.F. Acsadi, G. Johnson-Acsadi & R.A. Bulatao (eds.), *Population Growth and Reproduction in Sub-Saharan Africa: Technical Analyses of Fertility and Its Consequences*. Washington, DC: World Bank, 3.
21. Ibid., 43.
22. See, for example, Brown, L.R. & Wolf, E.C., 1985: *Reversing Africa's Decline*. Worldwatch Paper No. 65. Washington, DC: Worldwatch Institute, *passim*.
23. Garcia, op. cit., 185.

Somalia

1. Janzen, J., 1990: Somalia. *Encyclopedia Britannica*. Eastern Africa, **17**, 838–46.
2. Krokfors, C., 1984: Environmental considerations and planning in Somalia. In: Labahn, T. (ed.), *Proceedings of the Second International*

Congress of Somali Studies, August 1–6 1993. Vol. III, *Aspects of Development*. Hamburg: University of Hamburg, 293–312.
3. Janzen, J., 1984: Nomadismus in Somalia. Struktur der Wanderweidewirtschaft und Hintergränd aktueller Entwicklungsprobleme im monadischen Lebensraum-eimberblock. *Afrika Spectrum*, **19**, 149–71; Swift, J., 1977: Pastoral development in Somalia: herding cooperatives as a strategy against desertification and famine. In: M.H. Glantz (ed.), *Desertification: Environmental Degradation in and around Arid Lands*. Boulder, Colorado: Westview Press, 275–305.
4. Janzen, J., 1986: Ländliche Entwicklung in Somalia: Strukturen, Probleme, Tendenzen. *Geographische Rundschau*, **38**, 11, 557–64.
5. Janzen, J., 1984: op. cit.; Janzen, J., 1986: The process of nomatid sedentarisation: distinguishing features, problems and consequences for Somali development policy. In: Conze, P. & Laban, T. (eds.), *Somalia: Agriculture in the Winds of Change*. Hamburg: University of Hamburg, EPI Dokumentation, 73–91; Krokfors, op. cit.
6. See, for example, Glantz, M.H., 1976: Water and inappropriate technology: deep wells in the Sahel. *Journal of International Law and Policy*, **6**, 527–40.

The Brazilian Nordeste

1. Magalhães, A.R. & Glantz, M.H. (eds.), 1992: *Socioeconomic Impacts of Climate Variations and Policy Responses in Brazil*. Brasilia: Esquel Brazil Foundation.
2. Roett, R., 1972: *The Politics of Foreign Aid in the Brazilian Northeast*. Nashville, Tennessee: Vanderbilt University Press, 11.
3. Kutcher, G.P. & Scandizzo, P.L., 1981: *The Agricultural Economy of Northeast Brazil*. World Bank Research Publication. Baltimore, Maryland: Johns Hopkins University Press, 6.
4. Magalhães, A.R. & Rebouças, O.E., 1988: Introduction: drought as a policy and planning issue in Northeast Brazil. In: M.L. Parry, T.R. Carter & N.T. Konijn (eds.), *The Impact of Climatic Variations on Agriculture*. Vol. 2, *Assessments in Semiarid Regions*. Dordrecht, The Netherlands: Kluwer Academic Publishers, 279–304.
5. Kutcher and Scandizzo, op. cit., 33.
6. Earthscan, 1984: *Environment and Conflict*. Earthscan Briefing Document No. 40. London, UK: International Institute for Environment and Development, 51.
7. Greenland, D.J., 1975: Bringing the Green Revolution to the shifting cultivator: better seed, fertilizers, zero or minimum tillage, and mixed cropping are necessary. *Science*, **190**, 4217, 28 November, 841–4.

8. Magalhães and Glantz, op. cit.
9. Boulding, K., 1964: *The Meaning of the Twentieth Century: The Great Transition.* New York: Harper & Row.

The dry regions of Kenya

1. The author gratefully acknowledges the excellent comments and suggestions of Ted Bernard, Professor of Geography, Ohio University, Athens, Ohio.
2. Trewartha, G.T., 1981: *The Earth's Problem Climates,* 2nd ed. Madison, Wisconsin: University of Wisconsin Press, 134.
3. Ogallo, L. & Nassib, I.R., 1984: Droughts and famines in East Africa. Second Symposium on Tropical Droughts. *WMO Tropical Meteorological Program Report,* **15,** 41–4; Downing, T.E., Gitu, K.W. & Kamau, C.M. (eds.), 1989: *Coping with Drought in Kenya: National and Local Strategies.* Boulder, Colorado: Lynne Rienner Publishers.
4. Mott, F.L. & Mott, S.H., 1980: Kenya's population growth: a dilemma of development. *Population Bulletin,* **35,** 3, 1–35.
5. World Bank, 1992. *World Development Report: Development in the Environment.* New York: Oxford University Press.
6. Downing *et al.,* op. cit.
7. Miller, N.N., 1984: *Kenya: The Quest for Prosperity.* Boulder, Colorado: Westview Press.
8. Kitching, G., 1980: *Class and Economic Change in Kenya: The Making of an African Petite-Bourgeoisie, 1905–1970.* New Haven: Yale University Press.
9. Greenland, D.J., 1975: Bringing the Green Revolution to the shifting cultivator: better seed, fertilizers, zero or minimum tillage, and mixed cropping are necessary. *Science,* **190,** 4217, 28 November, 842.
10. Wasserman, G., 1976: *Politics of Decolonization: Kenya Europeans and the Land Use Issue 1960–1965.* Cambridge: Cambridge University Press.
11. Lofchie, M.F., 1990: Kenya: still an economic miracle? *Current History,* **89,** 211.
12. Ibid.
13. Bernard, F.E., Campbell, D.J. & Thom, D.J., 1989: Carrying capacity of the eastern ecological gradient of Kenya. *National Geographic Research,* **5,** 4, 399.
14. Campbell, D.J., 1981: Land use competition at the margins of the rangelands: an issue in development strategies for semi-arid areas. In: G. Norcliffe & T. Pinfold (eds.), *Planning African Development.* Boulder, Colorado: Westview Press, 49.

15. Little, P.D., Horowitz, M.M. & Nyerges, A.E. (eds.), 1987: *Lands at Risk in the Third World: Local-level Perspectives*. Boulder, Colorado: Westview Press, 199.
16. Kenya, 1921: *Annual Report: Maasai District*. Nairobi, Kenya: District Office.
17. Campbell, op. cit.; Little *et al.*, op. cit.; Western, D. 1982: Amboseli National Park: enlisting landowners to conserve migratory wildlife. *Ambio*, **11**, 302–8.
18. Western, op. cit.
19. Campbell, D.J. & Olson, J.M., 1991: Environment and development in Kenya: flying the kite in Kajiado District. *The Centennial Review*, **35**, 2, 310.
20. Ibid., 311.
21. Norse, D., 1992: A new strategy for feeding a crowded planet. *Environment*, **34**, 6–11, 32.
22. Campbell & Olson, op. cit.
23. Ibid.

Australia

1. Heathcote, R.L., 1969: Drought in Australia: a problem of perception. *Geographical Review*, **59**, 175–94; Heathcote, R.L., 1988: Drought in Australia: still a problem of perception. *GeoJournal*, **16**, 4, 387–97.
2. Campbell, R., Crowley, P. & Demura, P., 1983: Impact of drought on national income and employment. *Quarterly Review of Rural Economy*, **5**, 3, 254–7.
3. White, G.F., 1974: *Natural Hazards: Local, National, Global*. New York: Oxford University Press.
4. Foley, J.C., 1957: Drought in Australia. *Commonwealth of Australia Bureau of Meteorology*, Bulletin No. 43. Melbourne, Australia: Bureau of Meteorology; Gibbs, W.J. & Maher, J.V., 1967: Rainfall deciles as drought indicators. *Commonwealth of Australia Bureau of Meteorology*, Bulletin No. 48. Melbourne, Australia: Bureau of Meteorology.
5. Dijik, M. van, Mercer, D. & Peterson, J., 1983: Australia's drought and the southern climate. *New Scientist*, 7 April, 30–2.
6. Heathcote, R.L., 1986: Drought mitigation in Australia: reducing the losses but not removing the hazard. *Great Plains Quarterly*, **6**, 3, 225–37; Heathcote, R.L., 1991: Managing the drought? Perception of resource management in the face of the drought hazard in Australia. *Vegetatio*, **91**, 219–30.
7. Meinig, D.W., 1962: *On the Margins of the Good Earth: The South Australian Wheat Frontier, 1869–1884*. Chicago, Illinois: Rand McNally.

8. Heathcote, R.L., 1980: Perception of desertification in the Murray Mallee of southern Australia. In: R.L. Heathcote (ed.), *Perception of Desertification*. Tokyo, Japan: United Nations University, 60–96; Heathcote, R.L., 1991: Probleme der Besiedlung des Murray-Mallee-Gebietes (Südaustralien). *Geographische Rundschau*, **43**, 454–60.
9. Trumble, H.C., 1948: Rainfall, evaporation and drought-frequency in South Australia. *Journal of Agriculture*, **52**, 55–64, and Supplement, 1–15.
10. Cornish, E.A., French, R.A. & Hill, G.W., 1980: Yield trends in the wheat belt of South Australia from 1896 to 1964. *Agricultural Record*, **7**, 12, 3–32; Smailes, P. & Heathcote, R.L., 1992: Holding the line or chasing the rainbow? The quest for sustainability on Eyre Peninsula, South Australia. In: A.W. Gilg *et al.* (eds.), *Progress in Rural Policy and Planning*, Vol. 2. London: Belhaven Press, 234–46.
11. Anderson, J.A., 1979: Impacts of climatic variability in Australian agriculture: a review. *Review of Marketing and Agricultural Economics*, **47**, 147–78.
12. Heathcote, R.L., 1965: *Back of Bourke: A Study of Land Appraisal and Settlement in Semi-Arid Australia*. Carlton, Australia: Melbourne University Press, 158.
13. Young, M., 1979: Pressures and constraints on arid land management. Second Conference of the Australia Rangeland Society. Adelaide. Mimeo.
14. Cornish *et al.*, op. cit.
15. Smailes & Heathcote, op. cit.
16. Gentilli, J., 1971: Climates of Australia. *World Survey of Climatology*, **13**, 35–211. Amsterdam: Elsevier.
17. Pittock, A.B., 1981: Long-term climatic trends in eastern Australia. Proceedings of Seminar on Cropping on the Margins. Australian Institute of Agricultural Science and the Water Research Foundation. Mimeo.
18. Proctor, M.L.R., 1940: Marginal land: South Australia and New South Wales compared. *Australian Geographer*, **3**, 8, 16–31.
19. Heathcote, R.L., 1975: *Australia*. London: Longman.
20. Harte, A.J., 1984: Effect of tillage on the stability of three red soils of the northern wheat belt. *Journal Soil Conservation Service New South Wales*, **40**, 94–101.
21. Smith, D.I. & Callahan, S.D., 1988: Climatic and agricultural drought, payments and policy: a study of New South Wales. CRES Working Paper 88–16. Canberra, Australia: Australian National University.

22. DPRTF, 1990: *National Drought Policy Final Report*, Vol. 1. Canberra, Australia: Australian Government Publishing Service, 3.

Ethiopia

1. Sen, A., 1981: *Poverty and Famines*. London: Oxford University Press.
2. Gryseels, G. & Anderson, F., 1983: *Research on Farm and Livestock Productivity in the Central Ethiopian Highlands: Initial Results 1977–1980*. Addis Ababa, Ethiopia: International Livestock Centre for Africa.
3. McCann, J., 1986: Households, peasants, and the push factor in northern Ethiopian history. *Review*, winter issue, 369–411.
4. Sillani, T., 1933: *L'Africa Orientale Italiana (Eritrea e Somalia)*. Rome, Italy; Taddia, I., 1986: *L'Eritrea-Colonia 1890–1952: Paesaggi, Strutture, Uomini del Colonialismo*. Milan, Italy: Franco Angeli; McCann, J., 1987a: Unpublished report to Oxfam, UK, on the Evaluation of Hararge Projects; McCann, J., 1987b: *From Poverty to Famine in Northeast Ethiopia: A Rural History*. Philadelphia, Pennsylvania: University of Pennsylvania Press; McCann, J., 1987c: The social impact of drought in Ethiopia: oxen, households, and some implications for rehabilitation. In: M.H. Glantz (ed.), *Drought and Hunger in Africa: Denying Famine a Future*. Cambridge: Cambridge University Press, 245–68.
5. Tekeste, N., 1988: *Italian Colonialism in Eritrea, 1882–1941: Policies, Praxis and Impact*. Uppsala: Almqvist and Wiksell; McCann, 1987b, op. cit.
6. McCann, 1987b, op. cit., 187.
7. Messerli, B. & Aerni, K. (eds.), 1978: *Simen Mountains, Ethiopia*. Vol. 1, *Cartography and Its Application for Geographical and Ecological Problems*. Bern: Geographische Institut de Universität Bern.
8. McCann, 1987a, op. cit.
9. Ege, S., 1978: *Chiefs and Peasants: The Socio-Political Structure of the Kingdom of Shawa about 1840*. MA Thesis, University of Bergen.
10. McCann, 1987a, op. cit.
11. Gryseels & Anderson, op. cit.
12. Kloos, H., 1982: Development, drought and famine in the Awash Valley of Ethiopia. *African Studies Review*, December, 21–48.
13. Langdon, J., 1986: *Horses, Oxen, and Technological Innovation: The Use of Draft Animals in English Farming from 1066 to 1500*. Cambridge: Cambridge University Press; Ladurie, E.L., 1972: *Peasants of Languedoc*. Urbana, Illinois: University of Illinois Press.
14. Dessalegn, R., 1984: *Agrarian Reform in Ethiopia*. Uppsala: School of Advanced International Studies (SAIS), Johns Hopkins University.

15. Merid Wolde Aregay, 1986: Land tenure and agricultural productivity, 1500–1850. *Proceedings of the Third Annual Seminar of the Department of History.* Addis Ababa, Ethiopia: University of Addis Ababa.
16. McCann, 1987c, op. cit.
17. Ibid.
18. Dessalegn, op. cit.

Northwest Africa

1. Bowden, L., 1979: Development of present dryland farming systems. In: A.E. Hall, G.H. Cannell & H.W. Lawton (eds.), *Agriculture in Semi-Arid Environments.* Berlin: Springer-Verlag, 61.
2. Montagne, R., 1952: Naissance et développement du prolétariat marocain. In: C. Celier *et al.*, *Industrialisation de l'Afrique du Nord.* Paris, France: A. Colin, 199–222.
3. For example, see Swearingen, W., 1987: *Moroccan Mirages: Agrarian Dreams and Deceptions, 1912–1986.* Princeton, New Jersey: Princeton University Press, especially 8–10 and 193–196; Shaw, T., 1981: Climate, environment and history: the case of Roman North Africa. In: T.M.L. Wigley, M.J. Ingram & G. Farmer (eds.), *Climate and History: Studies in Past Climates and Their Impact on Man.* Cambridge: Cambridge University Press, 379–403.
4. Dresch, J., 1956: Remarques sur l'homme et al dégradation des écosystèmes naturels au Maghreb. *La Pensée*, **252**, 89–95.
5. Cleaver, K., 1982: *The Agricultural Development Experience of Algeria, Morocco, and Tunisia: A Comparison of Strategies for Growth.* World Bank Working Paper No. 552. Washington, DC: World Bank.

The Virgin Lands Scheme in the former Soviet Union

1. Lydolph, P.E., 1979: *Geography of the U.S.S.R.: Topical Analysis.* Elkhart Lake, Wisconsin: Misty Valley Publishing, 222.
2. McCauley, M., 1976: *Khrushchev and the Development of Soviet Agriculture.* New York: Holmes & Meier Publishers, xii.
3. As quoted in Kuznetsov, A.T., 1959: Climatic zones. In: *Climate of Kazakhstan.* Leningrad: L. Gidrometeoizdat, 76–7.
4. Hutchinson, C.F., Warshall, P., Arnould, E.J. & Kindler, J., 1992: Development in arid lands: lessons from Lake Chad. *Environment*, **34**, 6, 41.
5. Baishev, S.B., 1979: The historical significance of opening up the virgin and idle lands. *Voprosy Ekonomiki*, **3**, 52–8.
6. McCauley, op. cit.

7. Maslov, N., 1980: Virgin lands: 25 years later. *Soviet Life*, May, 6.
8. McCauley, op. cit.
9. Crankshaw, E., 1966: *Khrushchev: A Career*. New York: Viking Press, 179–80.
10. McCauley, op. cit.
11. Smith, G.A.E., 1987: The Stalinist legacy and the pressure for reform. In: M. McCauley (ed.), *Khrushchev and Khrushchevism*. Bloomington, Indiana: Indiana University Press, 103.
12. Ibid., 103–4.
13. Wheeler, G., 1964: *The Modern History of Soviet Central Asia*. Westport, Connecticut: Greenwood Press, 162.
14. McCauley, op. cit.
15. Brezhnev, L., 1979: *The Virgin Lands*. Moscow: Progress Publishers.
16. McCauley, op. cit., 90–7.
17. Pilatov, P.N., 1966: *Steppes of the USSR as a Condition of the Material Life of Society*. USSR: Yaroslavl State University, 118.
18. Brezhnev, op. cit.
19. Olcott, M.G., 1987: *The Kazakhs*. Stanford, California: Hoover Institution Press, 237.

South Africa

1. Abrams, L., Short, R. & Evans, J., 1992: *Root Cause and Relief Restraint Report*, Consultative Forum on Drought. Secretarial and Ops Room, Johannesburg, 8 October.
2. See Part I, this volume, 18.
3. Van Rooyen, J., Vink, N. & Malatsi, M., 1992: Agricultural change, the farm sector and the land issue in South Africa. Occasional Papers, Konrad-Adenhauer-Stiftung, Johannesburg, 5.
4. Kirsten, J. & Van Zyl, J., 1993: Agriculture and food security in South Africa. Paper presented at a conference on food security in South Africa, hosted by the National Consultative Forum on Drought, Johannesburg, South Africa, 14–15 June; Lenta, G., 1986: Food production in the homelands: constraints and remedial policies. RSA 2000, **8**, 1, 36–41.
5. Brand, S.S., Christodoulou, N.T., Van Rooyen, C.J. & Vink, N., 1991: Agriculture and redistribution: a growth with equity approach. Unpublished paper, Development Bank of Southern Africa.
6. Lindesay, J.A., 1993: Present climates of southern Africa. In: J.E. Hobbs (ed.), *Southern Hemisphere Climates: Present, Past and Future*. London: Belhaven Press (in press).

7. Scotney, D.M., Volschenk, J.E. & Van Heerden, P.S., 1990: The potential and utilization of the natural resources of South Africa. Department of Agricultural Development, 1–11.
8. Van Zyl, J., Van der Vyver, A. & Groenewald, J.A., 1987: The influence of drought and general economic effects on agriculture: a macro-analysis. *Agrekon*, **2**, 1, 8–12.
9. Brand *et al.*, op. cit.
10. Scotney, D.M., 1988: The agricultural areas of Southern Africa. In: I.A.W. MacDonald & R.J.M. Crawford (eds.), *Long-term Data Series Relating to Southern Africa's Renewable Natural Resources*. South African National Scientific Programmes, Report Number 157, 316–36.
11. Schoeman, J.L. & Scotney, D.M., 1987: Agricultural potential as determined by soil, terrain and climate. *South African Journal of Science*, **83**, 260–8; Cooper, D., 1988: Working the land: a review of agriculture in South Africa. Environment and Development Agency.
12. Cooper, D., 1983: Land reform and rural development in the Transvaal. *Development Southern Africa*, **4**, 492–500.
13. Schoeman & Scotney, op. cit., 267.
14. Schoeman & Scotney, op. cit.; Cooper, 1988, op. cit.
15. Cooper, 1988, op. cit.
16. Tapson, D.R., 1985: The agricultural potential of the homelands: problems and prospects. In: H. Gilomee & L. Schlemmer (eds.), *Up Against the Fences*. Cape Town: David Philip, 234–41.
17. McKenzie, C.C., Weiner, D. & Vink, N., 1989: Land use, agricultural productivity and farming systems in Southern Africa. Development Bank of Southern Africa.
18. Cooper, 1988, op. cit.; Cooper, 1983, op. cit.; Trollope, W.S.W., 1985: Third world challenges for pasture scientists in southern Africa. *Journal of the Grassland Society of Southern Africa*, **2**, 14–17; Tapson, D.R., 1990: Rural development and the homelands. *Development Southern Africa*, **7**, 561–81; Lynne, M.C. & Niewoudt, W.L., 1991: Inefficient land use in Kwa Zulu: causes and remedies. *Development Southern Africa*, **8**, 193–201; Bekker, S., Cross, C. & Bromberger, N., 1992: *The Wretched of the Earth*. Indicator South Africa, Centre for Applied Social Studies, Durban, 53–60.
19. Brand *et al.*, op. cit.
20. Marcus, T., 1989: Modernizing super-exploitation: restructuring South African agriculture. Dr Govan Mbeki Fund, Amsterdam: University of Amsterdam; Brand *et al.*, op. cit.; Van Zyl, J. & Vink, N., 1992: The mini-farmer approach: a case study of the South African tea industry. *Development Southern Africa*, **9**, 365–80.

21. National Maize Producers' Organization (NAMPO), 1993: *Maize Production in the Nineteen Nineties.* Bloemfontein: Dreyer Publishers.
22. Van Rooyen, J., Vink, N. & Malatsi, M., 1992: Agricultural change, the farm sector and the land issue in South Africa. Occasional Papers, Johannesburg: Konrad-Adenhauer-Stiftung.
23. Ibid.
24. van Zyl & Vink, op. cit.
25. Ibid.
26. Mather, C., 1992: Beyond the farm: rural and agricultural geography. In: C. Rogerson & J. McCarthy (eds.), *Geography in a Changing South Africa: Progress and Prospects.* Cape Town: Oxford University Press, 229–45.
27. Brand *et al.*, op. cit.
28. van Zyl & Vink, op. cit.
29. Brand *et al.*, op. cit.; De Klerk, M. (ed.), 1991: *A Harvest of Discontent: The Land Question in South Africa.* Institute for Democratic Alternative for South Africa, Hill House, Cape Town; Mather, op. cit.; van Zyl & Vink, op. cit.
30. van Zyl & Vink, op. cit.
31. Ibid.
32. Mather, op. cit.
33. Ibid.
34. Brand *et al.*, op. cit.
35. Lipton, M., 1977: South Africa, two agricultures. In: F. Wilson, A. Kooy & D. Hendrie (eds.), *Farm Labour in South Africa.* Cape Town: David Philip, 72–85; Nattrass, J., 1981: *The South African Economy.* Cape Town: Oxford University Press.
36. Van Rooyen *et al.*, op. cit.
37. Beinart, W., 1982: *The Political Economy of Pondoland, 1865–1930.* Cambridge: Cambridge University Press; Bundy, C., 1988: *The Rise and Fall of the South African Peasantry*, 2nd ed. Cape Town: David Philip.
38. Lenta, op. cit.
39. Daphne, P., 1982: *Tribal Authority and Community Organisation.* Occasional Papers 3, Centre for Research and Documentation, University of Zululand; Cooper, 1983, op. cit.
40. Lenta, op. cit.
41. Low, A., 1986: *Agricultural Development in Southern Africa: Farm Household Economics and the Food Crisis.* London: James Currey.
42. McKenzie *et al.*, op. cit.; Wilson, F. & Ramphele, M., 1989: *Uprooting Poverty: The South African Challenge.* Cape Town: David Philip.
43. de Wet, C., 1987: The impact of environmental changes on relocated

communities in South Africa. *Development Southern Africa*, **4**, 312–23.
44. de Wet, C., 1987, op. cit.; de Wet, C., 1989: Betterment planning in a rural village in Keiskammahoek, Ciskei. *Journal of Southern African Studies*, **15**, 326–45; Cooper, op. cit.; Bekker *et al.*, op. cit.
45. de Wet, 1989, op. cit.
46. de Wet, 1987, op. cit.; de Wet, 1989, op. cit.; Cooper, 1988, op. cit.
47. Cooper, 1988, op. cit., 94.
48. Erskine, J.M., 1983: Impact of drought in Natal/Kwa Zulu. *South African Journal of Science*, **19**, 439–40; de Wet, 1987, op. cit.; de Wet, 1989, op. cit.; Cooper, 1988, op. cit.
49. Wilson, F., 1991: A land out of balance. In: M. Ramphele (ed.), *Restoring the Land: Environment and Change in Post-apartheid South Africa*, 27–38. London: Panos Publishing.
50. Union of South Africa, 1923: *Final Report of the Drought Investigation Commission*. Cape Times Limited, Government Printers, 5.
51. Verbeek, W.A., 1965: *Report on Drought Feeding*. Department of Agricultural Technical Services, Government Printers, Pretoria, 7, paragraph 26.
52. Republic of South Africa, 1968: *Interim Report of the Commission of Inquiry into Agriculture*. RP 61, Government Printer, Pretoria.
53. Verbeek, op. cit., 14.
54. Republic of South Africa, 1972: *Final Report of the Commission of Inquiry into Agriculture*. RP 84, Government Printer, Pretoria, paragraph 10.1.1.3.
55. Cooper, 1988, op. cit.
56. Ibid.
57. Lenta, op. cit.
58. Bolus, R. & Miller, N., 1984: The drought and underdevelopment in the Transkei, 1982–83; *South African Review II*, South African Research Services, Ravan Press, 290–9; Bembridge, T.J., 1987: Crop farming system constraints in Transkei: implications for research and extension. *Development Southern Africa*, **4**, 67–89.
59. Forbes, R.G. & Trollope, W.S.W., 1991: Veld management in the communal areas of Ciskei. *Journal of the Grassland Society of Southern Africa*, **8**, 147–50.
60. Fenyes, T.I. & Groenewald, J.A., 1985: Food production in Lebowa: the interaction of social, physical and economic considerations. *South African Journal of Agricultural Extension*, **14**, 46–56.
61. Erskine, op. cit.
62. Freeman, C., 1988: The changing impact of drought on rural Tswana society. Unpublished MA dissertation, University of the Witwatersrand; Vogel, C., 1994: Consequences of droughts in southern

Africa (1960–1992). PhD thesis, University of the Witwatersrand (in preparation); Vogel, C. & Drummond, J., 1993: Dimensions of drought: South African case studies. *Geojournal*, **30**, 85–98.
63. Bembridge, op. cit.; Bolus & Miller, op. cit.
64. Bembridge, op. cit.
65. Bolus & Miller, op. cit.
66. Beinart, op. cit.
67. Bembridge, op. cit.
68. Ibid.
69. Bolus & Miller, op. cit., 298.
70. De Klerk, C.H., 1986: 'n Ondersoek na faktore wat in die weg staan van die aanvaarding van aanbevolle veldbeheerpraktyke (An investigation into the factors that prevent the acceptance of recommended grassland management). Pretoria: Dept. van Landbou en Waternavoorsiening.
71. Department of Agriculture, 1983: Droogte verslag van die seisoen (Drought report of the season). Pretoria: Department of Agriculture.
72. Du Toit, P.F., Aucamp, A.J. & Bruwer, J.J., 1991: The national grazing strategy of the Republic of South Africa: objectives, achievements and future challenges. *Journal of the Grassland Society of Southern Africa*, **8**, 126–30.
73. NAMPO, op. cit.
74. Van Zyl, J. & De Jager, F.J., 1991: 'n Ekonomiese ontleding van die grondontwikkelingskema in Wes-Transvaal (An economic analysis of the land conversion scheme in the Western Transvaal). *Agrekon*, **30**, 143–5; NAMPO, op. cit.
75. Department of Agricultural Development, 1992: *Assistance to Agriculture, Schemes for Implementation*.
76. Ibid.
77. Hobson, S. & Short, R., 1993: A perspective on the 1991–92 drought in South Africa. *Drought Network News*, **5**, 3–6.
78. Abrams, L., Short, R. & Evans, J., 1992: Root cause and relief restraint report, Consultative Forum on Drought. Secretarial and Ops Room, Johannesburg, 8 October; Hobson & Short, op. cit.; Vogel & Drummond, 1993.
79. Abrams *et al.*, op. cit.
80. Byford-Jones, C., 1989: Agroforestry: land-saver and profit maker. *Farmer's Weekly*. 19 May, 8–10; Cooper, D., 1983: Looking at development projects. *Work in Progress*, **26**, 29–37; Pretorius, L., 1987: Marginale mieliegrond word goudmyn (Marginal maize land becomes a gold mine). *Landbouweekblad* (Afrikaans version of *Farmer's Weekly*), 6 March, 20–3; Martin, F., 1988: Spectacular maize yields. *Farmer's Weekly*, 8 January, 6–9.

Is the stork outrunning the plow?

1. Mooneyham, W.S., 1975: *What Do You Say to a Hungry World?* Waco, Texas: Word Book Publishers.
2. Meadows, D.H., Meadows, D.L., Randers, J. & Behrens, W.W. III, 1972: *The Limits to Growth.* New York: Universe; Mesarovic, M. & Pestel, E., 1974: *Mankind at the Turning Point.* New York: Dutton.
3. Maddox, J., 1972: *The Doomsday Syndrome.* New York: McGraw-Hill.
4. WCED (World Commission on Environment and Development), 1987: *Our Common Future.* New York: Oxford University Press.
5. Boulding, K., 1964: *The Meaning of the Twentieth Century: The Great Transition.* New York: Harper & Row, 126–7.
6. Kates, R.W. & Haarman, V., 1992: Where the poor live. *Environment,* **34,** 4–11, 25–8.
7. Anon., 1993: The silk road catches fire. *The Economist,* 8 January, 44–6.

Index

Africa *see* Horn of Africa; Northwest Africa; *named states*
Albedo
  defined 14–15
  increase in 148
Amazon–Nordeste Brazil connection 73–6
Australia 91–102
  1982–3 drought effects 93, 101
  drought reporting inaccuracy 101
  New South Wales 96, 99
  overstocking (1895–1902) 96–8
  settlement, environmental impact 100–1
  South Australia 94–6
  wheat
    expansion 93, 94–100
    failed wheat frontier 95

Brazil, Nordeste
  case study 59–76
  colonial legacy 63–4
  drought policies 66–73
  drought polygon 61–3
  drought recurrence 61–3
  ecosystems 59–61
  extending agricultural frontiers 64–6
  land ownership 65–6
  Nordeste–Amazon connection 73–6
  SUDENE 59, 71
Brazil, Washington Accords 74
Brezhnev, Leonid, Virgin Lands Scheme, quoted 147

Canada, spring wheat region 10–11
Case studies, location 5
Cash crops 16, 98
  Sahel 42
Conservation cropping system 149

Desertification
  first use of term 12
  and food production 9–30
  Sahel 33–43
Drought
  defined 10
  Dhabaadheer Drought, Somalia 48
  and famine 11–12
  meteorological phenomenon 10
  reporting inaccuracy 101
  technological responses 91–3

Ethiopia 103–15
  1972–4 drought 42
  demography 107–10
  geography 103–7
    famine-prone areas 107
    marginal lands 104
  mean annual rainfall 105
  oxen supply and use 110–15
  population pressures 105, 108–9
  technology and social institutions 110–15

Famine (1992), and drought 11–12
Food production
  and desertification 9–30
  marginal lands 15–21

Index

Genesis Strategy, The 9
Global climate 175
Grain HYVs 15–17
Great Plains, US
  droughts (1930s and 1970s) 24, 26–30
  Dust Bowl 24, 28
  eighteenth century 1–2, 12, 22
  population shifts 24–5
  Texas aquifers 26–7
  tree-planting 23–4
Green Revolution, HYVs 15–17

Horn of Africa, rainfall 46–7

Kenya 77–89
  colonialism 85
  crops 83–5
  drought history 80
  land shortages 86–8
  Maasai 85
  Machakos district 88
  politics 82
  population distribution 81
  population growth 78
  rainfall 79
  wildlife policies 86–7
Khrushchev, Nikita, Virgin Lands Scheme 135–44
Kolumella, quoted 148

Maasai, Kenya 85
Malthusian prospects 172
Marginal lands
  defined 2
  food production 15–21
  reasons for extension of activities 172, 173–4

Northwest Africa 117–33
  agricultural policies 128
    marginal lands 130–2
    recent reforms 129–33
  agricultural zones 120
  climate variability 118–23
  colonization by Europeans 125–8
  defined 117
  environmental and social context of drought 118–24
  vulnerability to drought 124–8, 132–3

Opportunity cropping 98

Overstocking in Australia (1895–1902) 96–8

Population pressures 3–4
Population problem 171–5

Sahel
  1970s drought 13–15
  case study 33–43
  cash crops 42
  cultivators and pastoralists 18–19
  Sudan–Sahel region, climatic zones 34
Somalia
  case study 45–57
  desertification 51–2, 55
  Dhabaadheer Drought 48
  Jubba and Shabeelle rivers 50–1, 54
  northern 52
  political change 49–52
  recommendations 55
  regional land-use variation 52–7
  Sandridge belt 53
South Africa 151–70
  agriculture
    assistance measures 167–8
    in bantustans 153, 157, 160–1, 169
  Betterment Planning 160
  black rural areas 159, 163–4
  crop production potential 156
  diversity 150
  historical evidence 161–3
  Land Conversion Scheme 165–7
  National Grazing Strategy 164–5
  physical factors 152–7
  socioeconomic factors 157–61
  drought
    drought-stricken areas 154
    historical evidence 161–3
  livestock farming 163–4
  Marais–Du Plessis Commission 162
  population density 161
Soviet Union, former, Virgin Lands Scheme 19–20, 135–50
  grain belt 139
  initiation of soil degradation 144
  Kazakhstan 136–8, 140–1, 143
  State Plan for Remaking Nature 141–2
Stalin, State Plan for Remaking Nature 141–2
Sub-Saharan Africa *see* Sahel
Sudan, mechanization 21

Index

Sudan–Sahel region, climatic zones   34

Technology
  as a panacea   173
  water-related technologies   39–40
Texas aquifers, Great Plains   26–7
Tree-planting, Great Plains   23–4

Tuaregs   39

United States *see* Great Plains, US

Washington Accords, Brazil   74
West African Sahel *see* Sahel
Wheat, Australia   93

ACG-76M
11-20-95
20.0

VERMONT STATE COLLEGES
0 0003 0598948 5

DISCARD
LIBRARY
VT TECHNICAL COLLEGE
RANDOLPH CTR VT 05061